Which Oil?

Also from Veloce Publishing

Veloce's Essential Buyer's Guide Series
Alfa GT (Booker)
Alfa Romeo Spider Giulia(Booker & Talbott)
Audi TT (Davies)
Austin Seven (Barker)
Big Healeys (Trummel)
BMW E21 3 Series (1975-1983) (Reverente, Cook)
BMW GS (Henshaw)
BMW X5 (Saunders)
BSA 500 & 650 Twins (Henshaw)
BSA Bantam (Henshaw)
Citroën 2CV (Paxton)
Citroën ID & DS (Heilig)
Cobra Replicas (Ayre)
Corvette C2 Sting Ray 1963-1967 (Falconer)
Ducati Bevel Twins (Falloon)
Ducati Desmodue Twins (Falloon)
Ducati Desmoquattro Twins (Falloon)
Fiat 500 & 600 (Bobbitt)
Ford Capri (Paxton)
Ford Escort Mk1 & Mk2 (Williamson)
Ford Mustang (Cook)
Ford RS Cosworth Sierra & Escort (Williamson)
Harley-Davidson Big Twins (Henshaw)
Hinckley Triumph triples & fours 750, 900, 955, 1000, 1050, 1200 – 1991-2009 (Henshaw)
Honda CBR600 Hurricane (Henshaw)
Honda CBR FireBlade (Henshaw)
Honda SOHC fours 1969-1984 (Henshaw)

Jaguar E-type 3.8 & 4.2-litre (Crespin)
Jaguar E-type V12 5.3-litre (Crespin)
Jaguar XJ 1995-2003 (Crespin)
Jaguar XK8 & XKR (1996-2005) (Thorley)
Jaguar/Daimler XJ6, XJ12 & Sovereign (Crespin)
Jaguar/Daimler XJ40 (Crespin)
Jaguar Mark 1 & 2 (All models including Daimler 2.5-litre V8) 1955 to 1969 (Thorley)
Jaguar S-type – 1999 to 2007 (Thorley)
Jaguar X-type – 2001 to 2009 (Thorley)
Jaguar XJ-S (Crespin)
Jaugar XJ6, XJ8 & XJR (Thorley)
Jaguar XK 120, 140 & 150 (Thorley)
Kawasaki Z1 & Z900 (Orritt)
Land Rover Series I, II & IIA (Thurman)
Land Rover Series III (Thurman)
Lotus Seven replicas & Caterham 7: 1973-2013 (Hawkins)
Mazda MX-5 Miata (Mk1 1989-97 & Mk2 98-2001) (Crook)
Mercedes-Benz 280SL-560DSL Roadsters (Bass)
Mercedes-Benz 'Pagoda' 230SL, 250SL & 280SL (Bass)
MGA 1955-1962 (Sear, Crosier)
MGF & MG TF (Hawkins)
MGB & MGB GT (Williams)
MG Midget & A-H Sprite (Horler)
MG TD, TF & TF1500 (Jones)
Mini (Paxton)
Morris Minor & 1000 (Newell)

New Mini (Collins)
Norton Commando (Henshaw)
Peugeot 205 GTI (Blackburn)
Porsche 911 (930) Turbo series (Streather)
Porsche 911 (964) (Streather)
Porsche 911 (993) (Streather)
Porsche 911 (996) (Streather)
Porsche 911 Carrera 3.2 series 1984 to 1989 (Streather)
Porsche 911SC – Coupé, Targa, Cabriolet & RS Model years 1978-1983 (Streather)
Porsche 924 – All models 1976 to 1988 (Hodgkins)
Porsche 928 (Hemmings)
Porsche 930 Turbo & 911 (930) Turbo (Streather)
Porsche 944 (Higgins, Mitchell)
Porsche 986 Boxster series (Streather)
Porsche 987 Boxster and Cayman series (Streather)
Rolls-Royce Silver Shadow & Bentley T-Series (Bobbitt)
Subaru Impreza (Hobbs)
Triumph Bonneville (Henshaw)
Triumph Stag (Mort & Fox)
Triumph TR7 & TR8 (Williams)
Triumph Thunderbird, Trophy & Tiger (Henshaw)
Vespa Scooters – Classic two-stroke models 1960-2008 (Paxton)
Volvo 700/900 Series (Beavis)
VW Beetle (Cservenka & Copping)
VW Bus (Cservenka & Copping)
VW Golf GTI (Cservenka & Copping)

www.veloce.co.uk

For post publication news, updates and amendments relating to this book please visit www.veloce.co.uk/books/V4365

First published in August 2011, reprinted January 2014 by Veloce Publishing Limited, Veloce House, Parkway Farm Business Park, Middle Farm Way, Poundbury, Dorchester, Dorset, DT1 3AR, England.
Fax 01305 250479/e-mail info@veloce.co.uk/web www.veloce.co.uk or www.velocebooks.com.

ISBN: 978-1-845843-65-6 UPC: 6-36847-04365-0

© Richard Michell and Veloce Publishing 2011 & 2014. All rights reserved. With the exception of quoting brief passages for the purpose of review, no part of this publication may be recorded, reproduced or transmitted by any means, including photocopying, without the written permission of Veloce Publishing Ltd. Throughout this book logos, model names and designations, etc, have been used for the purposes of identification, illustration and decoration. Such names are the property of the trademark holder as this is not an official publication.
Readers with ideas for automotive books, or books on other transport or related hobby subjects, are invited to write to the editorial director of Veloce Publishing at the above address.
British Library Cataloguing in Publication Data – A catalogue record for this book is available from the British Library.
Typesetting, design and page make-up all by Veloce Publishing Ltd on Apple Mac. Printed in India by Replika Press.

Choosing the right oils & greases for your vintage, antique, classic or collector car

Which Oil?

Richard Michell

VELOCE PUBLISHING
THE PUBLISHER OF FINE AUTOMOTIVE BOOKS

Contents

1. Introduction & Acknowledgements 5
   Acknowledgements 6

2. Historical eras 7
   2.1 Prior to 1915 7
   2.2 1915-1929 9
   2.3 1930-1944 10
   2.4 1945-1959 11
   2.5 1960-1972 12
   2.6 1973-1979 14
   2.7 1980-1999 16
   2.8 Post 1999 20

3. Basics of lubrication 21
   3.1 Appropriate viscosity under the
       operating conditions 23
   3.2 Stability of the lubricant 31
   3.3 Protection of surfaces 34
   3.4 Protection of the lubricant delivery
       system 36
   3.5 Imperfect lubrication 37

4. The roles of lubricant components 40
   4.1 Reduction of oil thinning with
       temperature 40
   4.2 Reduction of oxidation
       susceptibility 45
   4.3 Dealing with mixed or partial
       lubrication regimes 46
   4.4 Dealing with boundary lubrication 47
   4.5 Dealing with emissions 49
   4.6 Friction modification 51

5. Engine lubrication 52
   5.1 Performance level 52
   5.2 Viscosity level 61
   5.3 Summary 62

6. Transmission lubrication (including
   final drive) 63
   6.1 Controlling the wear of the gear
       teeth 64
   6.2 Automatic transmissions 72

7. Chassis, steering and wheel bearing
   lubrication 76
   7.1 Types of grease 77
   7.2 Consistency of grease 79
   7.3 Grease performance specifications .. 80

8. Choosing appropriate lubricants
   for your car 81
   8.1 Engine 81
   8.2 Transmission 97
   8.3 Chassis and wheel bearings 109

9. Answers to some common
   questions 111
   9.1 Engine lubrication 111
   9.2 Manual gearboxes 120

10. Glossary 125

Index 126

1. Introduction & Acknowledgements

Proper lubrication is essential to the reliable operation and longevity of all vehicles – modern and classic – but what is proper lubrication? Unfortunately, while a small core of sound information is readily available, it is overlaid with – and often swamped by – a combination of mystery, myth, hype and hearsay.

For modern cars, manufacturers will give some guidance, although, if they sell so-called 'genuine' oils under their own brand, even they may not be very forthcoming. For an older car, even if the information that was published when the vehicle was new is available, the products that were specified or referred to most probably are not. The challenge for owners and drivers of such vehicles is to choose products from the modern lubricant range that will satisfy their vehicles' lubrication needs.

Proof that this challenge is real and of genuine concern to owners is shown very readily by even the most cursory search on the internet. Every car club site has a portion of its technical help site dominated by questions, comments and arguments over appropriate lubricants. Outside of these formal sites, there are dozens of other forums and discussions on this same topic.

The object of this book is to help you to successfully master this challenge. The approach taken is not to attempt to list every vehicle ever made and to give current lubricant recommendations. The list would be impossibly long and the recommendations would soon become out of date, as the modern lubricants evolved.

Rather, the book attempts to first give you essential underlying information and an understanding of the lubrication of the various mechanical components of motorcars, whatever their vintage. It then gives a suggested approach for you to follow – including some examples of its application – in order to select suitable modern lubricants. Hopefully, as a result you will be able to make sensible and safe selections for your vehicle from the lubricants that are available to you, allowing you to get on with the much more important business of enjoying the driving of your car.

The above describes the logical layout of the book. To get the most benefit from it you should read it in the order in which it is laid out. However, I recognise that the chapters that deal with the underlying lubrication concepts and needs are potentially heavy going and not necessarily highly exciting. I also suspect that, like me, you don't have infinite patience or time. Hence, many readers will no doubt turn straight to the last couple of chapters, the ones that explain how to make a lubricant choice and that deal with some of the common questions and points of contention. I confess that I probably would too.

If this is what you do, I would still encourage you to also read and digest at least part of the background information in earlier chapters. If you don't, then you

Which Oil?

may struggle with some of the terminology and concepts used. Also, the better your understanding is of the underlying principles, the better and safer will be your choice of lubricants and the less vulnerable you will be to the folklore and myths that abound.

Perhaps a good compromise, if you do start with the last chapters, is to consult the index or glossary of terms each time you strike a term or concept that you don't fully understand. Via the Index, go and read the relevant explanation. By doing this at the time that it is most relevant to your understanding, you'll be able to digest the background information in smaller, more manageable portions.

Irrespective of how you decide to approach the book, I trust that you will ultimately get practical benefit from it, along with new and interesting facts and some pleasure. To ease you in, and put you in the mood, in the next chapter I give a very potted outline of the development of the motor car over the course of the twentieth century, with particular emphasis on the evolution of its lubrication.

Acknowledgements

I am pleased and grateful to acknowledge the help of three particular people – David Blackwell, Alastair Browne, and Geoff Harrison. Each of them read the manuscript in its entirety, gave me invaluable advice, and, importantly, three different reading perspectives. The book is greatly enhanced through their efforts. However, even with their dedication, there will no doubt still be some errors. They are entirely my own responsibility.

2. Historical eras

There is an obvious difficulty in dividing cars into historical eras. At any given time, the vehicles being sold were not all of the same degree of technological development. So, while generalisations can be made, there will always be exceptions. You should always cross-check the information given in this chapter for the relevant period with what the manufacturer of your vehicle specified for its lubrication.

2.1 Prior to 1915

Prior to 1915, formulation of oils to lubricate engines was very much an art, rather than a science. There was limited knowledge and understanding, and so there was the opportunity for talented individuals and companies to make technological breakthroughs or advances. In this circumstance, what was critically important was the choice of the brand of lubricant. There was no such concept as a performance specification.

However, at the time that the motor car first emerged, there was already a significant knowledge of the lubrication of engines arising from their industrial use. Originally these industrial engines were steam driven, but internal combustion engines – often running on town or producer gas for their fuel – had become much more widespread by the start of the 1900s. However, these engines were, in general, large and slow revving when compared with those developed for transport and, obviously, they were stationary.

In this early period, the process of refining hydrocarbon or mineral oil (crude oil) was still very much in its embryonic state. Mineral oil-based products did not dominate lubricant technology as they do today. Products based on vegetable and animal oils, or on such oils mixed with mineral oil (known as compounded oils), were also widely used. The US tended to adopt mineral oil-based lubricants earlier than

---

Terminology

In this chapter, in order to recount the history that is relevant to the lubrication of motor cars, I am forced to use some technical terminology and concepts. These are explained fully in the following chapters, but at this stage they may not be totally clear to you.

If I am aware of it, why have I subjected you to this difficulty? My reason, paradoxical as it may seem, is to make the book easier and more enjoyable to read. By their nature, the chapters where the terms are explained require greater mental concentration than do the more general chapters, such as this one. Rather than dropping you straight in (please don't panic, there are only two hard ones – Chapters 3 and 4), I felt it preferable to ease you in by giving an historical overview first. If a word or concept puzzles you, I suggest that you look it up in the index and read enough of the text to which you are directed to alleviate your puzzlement.

Which Oil?

Europe. This was in part because of the large oil discoveries in the US, and in part because of an accident of circumstance (see following). Perhaps the best known of the compounded or mixed oils was Castrol R, a product based originally on castor oil (hence the Castrol name).

On the mineral oil side, it emerged that some crude oils gave better-performing lubricants than others. Crude mineral oils are classified as either paraffinic or naphthenic. In the paraffinic crude oils the hydrocarbon molecules that are present are mainly of a type that are called straight chain (see the panel). Less common, on a world scale, is the naphthenic type of crude. These crudes contain ring-structured hydrocarbon molecules (as well as a significant percentage of the straight chain type of molecules). Because of these differences in molecular make up, paraffinic oils thin out less with increases in temperature than naphthenic oils do. They are also more resistant to oxidation at high temperatures (reaction with oxygen from the air). However, because they contain paraffin waxes, they thicken more at low temperature than naphthenic oils do.

The modern-day hydrocarbon oil industry began commercially in Pennsylvania in the US in 1859. The deposits of crude oil found there were what we now know as paraffinic. As further oil was discovered in other areas of the US, and in Europe, particularly in Russia, it tended to be what we now call naphthenic. At the time the chemical differences were known but not fully understood. However, what was soon obvious was that lubricants made from Pennsylvanian crudes were superior in most circumstances and they gained a deservedly good reputation. In the US, to this day, a number of the lubricating oil marketers – outside of the big oil companies that also refine and sell fuel – had their origins in Pennsylvania and still have 'Pen' somewhere in their names.

One circumstance where the Pennsylvanian or paraffinic type of lubricants were inferior was at very low temperatures. As mentioned, they contain paraffin waxes which can solidify. The naphthenic crude-derived lubricants did not precipitate waxes and so were preferred in very cold climates.

Paraffinic and naphthenic crude oils

Crude mineral oils are made up of a large variety of hydrocarbon substances. As the name suggests, the molecules of a hydrocarbon contain mainly carbon atoms and hydrogen atoms – hence 'hydrocarbon.'

In some crude oils, the hydrocarbon molecules are mainly of a type that have the carbon atoms linked in a chain-like manner: -C-C-C-C-C-C-C-C-C-. These are known as straight chain molecules, and the types of crude oils that contain them are called paraffinic (after the substance paraffin or kerosine which is this type of hydrocarbon).

Hydrocarbon molecules can also form with the carbon atoms linked in a manner that forms a closed circle, or several joined circles. One particular form of circle gives rise to what are called aromatic molecules. Benzene is a well-known member of this family. Another form of closed or circular linking gives the molecules that are called naphthenic (the substance naphthene is a member of this group). Crude oils that contain a significant portion of such molecules are called naphthenic crudes.

Historical eras

In these early motor cars, engine oil consumption during use was very high. The primary concern was not drain interval but simply keeping the sump or oil reservoir (in drip feed systems) topped up during operation. It was not uncommon to use oil at a rate of as much as a pint every fifty miles or so (a litre every 100 to 150 kilometres). No chemical additives were used in the oils. They were straight mineral oils or mineral/vegetable oil mixtures.

Lubrication of transmissions and chassis was more straightforward than for the engines. In general, heavy oils – either mineral or compounded with natural products, again without any chemical additives – were used to lubricate gears, and animal fats were used for the chassis, although bitumastic materials were sometimes specified. Again, the US tended to be more mineral oil-based.

2.2 1915-1929

In this period, automotive design progressed very rapidly and started to converge. The strengths and weaknesses of the various approaches that different manufacturers had used in their attempts to solve the many individual design and operation problems were revealed through consumer experience. Mass production became the norm in the US and spread from there, further accelerating convergence of design and technology.

The understanding of lubrication and lubricants also advanced rapidly through the period, as did oil refining. Brand remained the only safe way to specify which lubricant to use, but the number of satisfactory brands increased. Lubricants became more consistent, and the performance differences between reputable brands lessened. By the end of the period, mineral oil-based lubricants had almost completely displaced animal or vegetable ones for automotive internal combustion engine use (although compounded oils lingered on in industrial applications). In the automotive area, engines were smaller, ran faster and hotter, and were generally more highly stressed.

The first use of chemical additives to enhance the performance of the base lubricating oil began in a limited way. Substances that lowered the temperature at which oils solidified due to wax precipitation were developed to extend the ability to use the more desirable paraffinic type oils in colder areas. Substances to reduce the foaming of oils when being churned around in engines and gearboxes/final drives were also introduced. Churning was much greater than in modern engines because, in general, oil systems were not fully pressurised. Lubrication was largely by splash and dip of the moving components into the oil in the sump.

Oil consumption during operation was reducing as machining tolerances and knowledge of materials and surface finishes improved, but it was still very high. For example, a high-performance six-cylinder Mercedes-Benz sports car from 1926 had an 8-litre sump, a 7.5-litre oil reservoir, and used about 15 litres (the entire oil capacity) every 1000 kilometres (600 miles). This was if the driver used a 'good' oil. A good oil was defined as one which was free of acids and had a low ash content. Three viscosity grades were specified for this engine – one each for Winter, Summer and Intermediate Seasons.

Which Oil?

This was a high-performance engine, and so its rate of oil consumption was at the high end of the typical range, but it still serves as a salutary reminder of the engineering advances that have been made. Recommended oil drain intervals varied quite widely between manufacturers but, by the end of the period they were typically of the order of 1000 miles (1500 kilometres).

Also, towards the end of the period, the SAE viscosity grading system – SAE 30, SAE 40, etc – was developed and introduced, in much the same form as we know it today.

The use of additives was not confined to the lubricant. Lead antiknock was introduced in the US in 1923 as an aftermarket additive, and, by 1925, leaded gasoline was widely available. From the US its use spread to Europe. The more consistent combustion that it gave allowed engine manufacturers to lift compression ratios and boost power. While this was a boon for the engine it was not a boon for the lubricating oil. As well as creating more demanding lubrication needs, the lead compounds – and particularly some of the other chemicals (called scavengers) that were used with them to reduce their tendency to form ash deposits in combustion chambers and on sparkplugs – formed acidic, corrosive substances which found their way into the engine oil.

Advances in gear oils – while not as dramatic as engine oils – were nevertheless significant. The first use of chemical additives to increase the oil's ability to resist heavy loads emerged. These are known as extreme pressure (EP) additives, and were generally based around sulphur, and often involved sperm whale oil and lead. However, mineral oil was becoming increasingly dominant as the prime basis of these oils.

2.3 1930-1944

By 1930, the concept of adding chemicals to the base lubricating oil to enhance its performance was much more widely applied. Engine oils with enhanced performance properties were introduced by all the leading oil marketing companies. They contained modest levels of chemical additives to help them resist being oxidised (reacting with oxygen in the air) at the high temperatures to which they were exposed in an engine, and to reduce the wear between rubbing metal surfaces.

The antioxidant additives were added to help extend the useful life of the oil. The desirability of this (at least by modern standards) is well shown by the fact that the 1931 Duesenberg Model J – perhaps the technology leader of its time, with a top speed of 116mph – had an engine-driven timing box of planetary gears that triggered indicator lights to (amongst other things) warn the driver to change the engine oil every 700 miles (1100 kilometres).

Oil refining techniques continued to advance and the base oil itself became more suitable for use as a lubricant, more resistant to thermal and oxidation change, and it thinned less at higher temperatures. A major development was solvent dewaxing. This greatly reduced the wax content of the desirable paraffinic base oils, allowing them to be used at lower temperatures.

Historical eras

The trend to add performance-enhancing chemicals to the base oil received a boost in the Second World War. During the war big advances were made in the science of lubricants, including the development of synthetic lubricants. Understandably, the developments were directed to military equipment and, from a technical perspective, the civilian market – for both lubricants and cars – largely stood still in the period 1939 to 1944.

Multigrade viscosity oils – oils which had chemical polymers added to them to further reduce their change in viscosity or thickness as the temperature changed – had not as yet been developed commercially, and so all engine oils specified were monogrades, with the consequent need for the user to change the viscosity grade used between winter and summer in many parts of the world. By the end of the period oil drain intervals had generally increased, although many manufacturers still recommended 1000 miles (1500km). More typical intervals were 2-3000 miles (3-4000km).

Transmission lubricants changed very little during the period but, in chassis applications, greases that were made by the chemical reaction of natural fats with manufactured compounds that contained calcium, displaced completely natural products, such as tallow. They were (and still are) called calcium greases for short.

2.4 1945-1959

Commercial oil technology (and car design) rapidly adopted the wartime technical developments, particularly in the US. As it advanced, the oil technology tended to converge, with the differences between brands lessening. This was driven to a significant extent by further advances in refining crude oil. The base oil for use in lubricants became less dependent on the geographical source of the crude – although paraffinic crudes were still the most desirable – and became of a higher and more consistent quality. However, brand remained very important in the marketplace, and many engine manufacturers made their lubricant recommendation by quoting particular products. In doing so they also specified – by means of the now widely-circulated SAE classification system – the appropriate viscosity grade to use in relation to the prevailing ambient temperature.

Car and engine design made even greater strides than oil technology in this period. Chassis construction largely disappeared, replaced by monocoque designs. Cars rode lower and bodywork was far more enveloping. This, combined with the greater power outputs, meant that oils ran hotter and were more highly stressed. The use of additives in the engine oils increased. The first full-flow oil filter system on a mass production vehicle was introduced in 1946, but most car engines throughout the period used partial systems.

As the period progressed, the use of standardised technical performance specifications for the engine lubricant became more widespread. This was in part because such specifications had become more sound and more widely publicised, but also because of the obvious and increasing limitation of the 'naming brands' approach, which was based on specific testing and approval by individual car manufacturers.

Which Oil?

By the end of the period oils with multigrade viscosity characteristics began to be available commercially, but they had only limited impact. The vast majority of consumers continued to use monogrades, with the consequent need to change the viscosity grade between winter and summer in many parts of the world. This was still not a great impost because recommended oil drain intervals were still only of the order of a few thousand miles and most owners had to change the oil at least twice a year in any case.

In this period, transmission and chassis lubricants also saw major changes. Mineral oil was now the universal base for these products, and extreme pressure (EP) and anti-wear additives became far more sophisticated as the loads in differentials increased. Also, the desire for lower vehicles lead to the widespread adoption of hypoid (offset) gears in these differentials. This geometry made the lubrication task even more difficult, requiring a further increase in the load carrying ability of their lubricants.

Performance specifications to define suitable gear oils for this application were developed and introduced, initially in the US but spreading rapidly to Europe and the rest of the world. Also in the US, automatic transmissions, with their somewhat specialised lubrication needs, grew from being almost nonexistent to dominant.

As vehicle speeds increased, the heat generated by the necessarily-improved brakes – often hydraulically actuated – required the adoption of higher melting point greases in the hubs and wheel bearings. Greases based on calcium or tallow were inadequate in this regard, and greases based on sodium became the standard wheel bearing lubricant, while calcium greases, with their high water resistance, continued in the chassis role.

2.5 1960-1972

As the 1960s dawned, the big development was the widespread commercial introduction of multigrade viscosity oils. Not all engine manufacturers embraced these oils initially and so there was not a universal recommendation for their use. There were several reasons for this (in addition to the fact that they were new and somewhat untried). One was that the polymers that were used to decrease the thinning of the base oil as its temperature rose, had a tendency to increase deposits on hot surfaces such as pistons. Detergent levels in the oil had to be increased to compensate. As technology evolved, polymers were developed that had their own detergent/dispersant (or 'keep clean') capability, and so this negative tendency was overcome. However, multigrade oils were (and still are) more expensive to produce than monogrades.

In Europe, particularly in the UK, the introduction of the Mini Minor in 1959 gave an enormous boost to multigrade oils. The gearbox in the Mini was in the sump of the engine and relied on the engine oil for its lubrication. To control gear wear, the oil had to be relatively viscous. However, a viscous oil gave difficulty with gear changes, particularly when the sump was cold. It also was not ideal for the lubrication needs of the engine. The solution was to use an SAE 20W/50 multigrade viscosity oil.

Because a gearbox is a high shear environment, if a multigrade oil was to perform

Historical eras

satisfactorily in service over time, the polymers used to formulate it had to be shear stable. With the technology available in 1960, the reality was that these SAE 20W/50 oils sheared down to a viscosity more like an SAE 15W/40 oil (or lower) in service, but this was still acceptable. A typical monograde oil recommended for engines at the time was SAE 30 or 40.

In the US the take up of multigrade viscosity engine oils was slower. There was a very strong tradition of seasonal oil changes in many parts of the US. Also the typical recommended maximum oil drain period was six months, and so changing the engine oil viscosity grade between winter and summer was not particularly onerous. Given that a multigrade oil was more expensive than a monograde, there was little obvious incentive for the consumer to change.

Unfortunately, these economic facts pushed the US oil industry towards the lowest cost route to multigrades – the use of low shear stability polymers. These oils were generally fine in typical American engines, which were of large capacity, low revving and under-stressed. However, some problems were experienced when they were used in higher performance imported European vehicles, or in more highly stressed US engines. As a result, multigrade oils developed a mixed reputation in the US. This was not the case in Europe where the Mini Minor – and the generally smaller, higher revving, more stressed engines typically used there – prevented any widespread introduction of low shear stability engine oils.

In this period the use of additives to boost the cleaning ability of engine oils – that is, their levels of detergency and, particularly, dispersancy – increased substantially. Part of the reason for this was to counter the deposit-forming tendency of the polymers added to reduce the thinning out of the oil with temperature (the viscosity index improvers), and part was because of the longer drain intervals that were evolving. In the US, performance in standardised multicylinder engine tests was adopted as the way to specify engine oils. Oils could not pass these tests without the addition of detergents to assist.

The engine tests were specified and overseen by a combination of motor and oil industry organisations. The outcome was called the American Petroleum Institute (API) Engine Oil Classification system. For motor cars, various levels of oil performance were defined, each level identified by a pair of letters. For cars, the first letter was 'S,' followed by 'A,' 'B,' etc. SA was the lowest performance level, the level increasing in the sequence SA, SB, etc. The system was applied retrospectively and, at the start of the period in which it was introduced, oil performance level was at the tail end of SB (actually called MM or MS at the time), but by the end of the period it had progressed through SC to SD.

With the introduction of dispersants (substances that helped keep solid particles in suspension) the design approach to oil filtration changed and the door was opened further to longer drain intervals. In the absence of dispersants, particles of soot (and other unwanted solid products of combustion) that found their way into the oil agglomerated in service. They were abrasive and could also settle out in oilways or on hot surfaces. Filtration could only remove them once they had grown to a size greater than that of the pores of the filter medium. However, a very fine

Which Oil?

filter could not be used because it would give too much pressure drop and restrict oil flow. In some designs this was overcome to some extent by using a fine filter, but passing only a portion of the oil through it. Obviously this was not an ideal solution either. Frequent draining of the oil was the only certain way to remove the particles.

Dispersants stopped the particles agglomerating and kept them in suspension in the oil. In this very fine state they were not nearly as detrimental to bearings and other surfaces. The job of the filter was now primarily to take out any larger particles (for example, dirt that entered the engine with the combustion air) and full flow filtration became the norm.

The arrival of dispersant oils did cause some confusion. These oils darkened more rapidly in service than the oils they superseded, because they kept the soot (which is black) in suspension. With non-dispersant oils, darkening had been viewed – quite rightly – as a sign that the oil itself was deteriorating. Oxidation products tend to be very dark. The rapid darkening of the new oils took some adjusting to.

Because dispersants reduced the need for oil drains to remove accumulating debris, the drain interval could be increased. However, this meant that the oil itself had to be able to survive the longer period of service. The use of oxidation inhibitor additives increased also. By the end of this period, oil drain intervals had grown to typically 5-6000 miles (8-10,000 kilometres).

Transmission designs did not change a great deal through this period. Automatics were now the dominant norm in the US but still relatively rare in Europe. Manual gearboxes became fully synchromesh, and many had five forward gears rather than the previous four (or three on larger-engined cars). Rear-wheel drive via a hypoid differential was the standard design, the Mini Minor and certain Citroëns being the main, large-volume exceptions.

The need for two greases – one for chassis and one for wheel bearings – was eliminated as lithium soap-based greases emerged. They were marketed as multipurpose greases and could satisfactorily handle all chassis and bearing lubrication tasks, being both high melting point and water-resistant.

2.6 1973-1979

1973 saw the arrival of the first oil crisis. Fuel prices took off and fuel economy, which had always been important in generally less affluent Europe, now became an issue in the US. Fuel economy targets were legislated for the American car industry (the CAFE legislation), and the car manufacturers suddenly became far more interested in engine oils and their viscosity. Multigrade oils, being less viscous than a monograde at low temperatures, but having the same or higher viscosity at high temperatures, could help. Any reservations the industry may have still had about recommending multigrade oils evaporated.

Also, car ownership was now virtually universal, with many families having two cars. The days when the building of freeways could keep up with the increasing demand were over, and traffic jams were now the norm in all major cities. The consequent stop/start driving put new demands on engines and their lubricants.

Historical eras

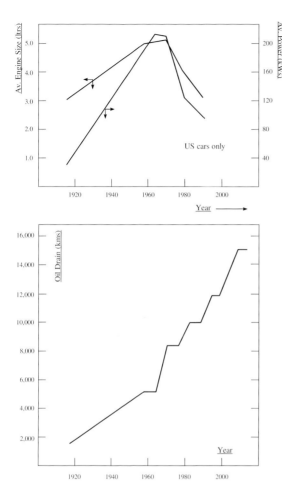

Fig 2.1 Change in engine power, size and oil drain interval (values in these two plots are average or typical figures).

The high level of car use, and the resulting traffic jams, also led to a deterioration of the air quality in cities due to emissions from car exhausts. Levels of tailpipe emissions had to meet new, more stringent, legislated requirements. Engine tune was altered and, in 1974 (US model year 1975), the first exhaust pipe catalysts appeared. So-called fugitive or evaporative emissions from fuel tanks and carburettors were also restricted, and the standard approach was to route these to the engine's combustion air inlet system. Crankcase ventilators were also connected to this inlet.

Over the period 1975 to 1986, lead was phased out of gasoline in much of the developed world, and octane lowered. This was partly for health reasons but also because the presence of lead would poison the exhaust catalysts. In most regions new cars sold after the period 1975 to 1986 had to run on unleaded petrol only.

Overall, with one major exception, the removal of lead from gasoline was beneficial from an engine lubrication perspective. The exception was wear between inlet and exhaust valves and their seats (valve recession), in those engines that did not have hardened valve seats. This was essentially all engines with cast iron cylinder heads. Lead in the gasoline, although not introduced for this purpose, had lubricated this contact area.

Which Oil?

Most of the other changes that occurred in engine design to meet emission legislation were negative and the overall outcome was a significant increase in the demands on the engine oil. The main industry engine oil specification was upgraded to SE in 1972, and this was a major performance hike over SD. Oil drain intervals continued to increase during this period, but not as steeply or as universally as in the previous one. By its end they were typically 5000-8000 miles (7500-12,000km).

A major change in Europe was the introduction, in 1976, of engine oil performance specifications, under the auspices of the CCMC (Committee of Common Market Automobile Constructors). These specifications were based on engine tests using a combination of US and European engines. Prior to 1976, European manufacturers had tended to use the US engine oil performance specifications. From 1976 they hedged their bets and specified both.

In the US, the cost of gasoline and the associated emphasis on fuel economy saw, for the first time in 60 years, the beginning of a significant and sustained reduction in car and engine size, and an associated reduction in the number of cylinders in the engine. The love affair with the V8 was ending. After an initial response that involved detuning and reduction of compression ratios, new designs began to emerge, and the stresses on all mechanical components began to increase, heading towards what had always been the European and Japanese approach. Technology was converging. Internationally, front-wheel drive – with the savings that it gave in the internal room in the downsized vehicles – began its rise to dominance.

In this period, transmission and chassis lubricants did not evolve a great deal. Some variations emerged in the lubricants specified by different manufacturers for manual gearboxes as monograde engine oils largely disappeared. The increasing adoption of disc brakes forced a further increase in the melting point of hub greases, with clay-based (bentone) greases emerging.

2.7 1980-1999

The pressures for fuel economy and low emissions continued and increased. On the emissions front, the automotive engineering tended to catch up with the problems and any remaining stop-gap measures – such as air pumps and exhaust gas recirculation – that had been adopted in the panic to meet emissions legislation, disappeared. Fuel-injection gradually replaced carburettors, and chokes became a thing of the past. This was good news for the lubricant. Soot loadings dropped and fuel dilution was no longer a significant factor.

On the other side of the ledger, because of the drive for fuel economy, engines in motor cars generally became smaller but had higher relative power outputs and more complex designs, such as overhead cams and multiple valves. For example, in the US between 1980 and 1988, average engine size reduced from 4.2 to 3.4 litres (256 to 207in^3) while compression ratio increased from 8.3 to 8.8. Turbocharging appeared on mass-produced engines for the first time.

The downsizing of cars meant that front-wheel drive, with its benefit of no need for an intrusive transmission tunnel and rear differential, became the norm. However,

Historical eras

unlike the earlier Mini Minor, the new designs for front-wheel drive kept the engine lubricant separated from that for the gearbox/final drive. With the downsizing, designs from the US and Europe started to converge, and this trend was accelerated by business-driven mergers. The Japanese influence grew massively as the market moved towards the size of car Japanese manufacturers specialised in, and Japanese models 'invaded' both the USA and Europe.

Towards the end of this period, aerodynamics became a significant factor in car design. This was because it had such a large influence on the engine power that was needed to reach a given top speed (and hence on the fuel consumption). However, the more effective diversion of air around the car tended to mean that engines and other components were less effectively cooled. They tended to operate at higher temperatures.

Also, even though cars were smaller, the level of accessories and creature comforts did not reduce. Air-conditioning remained a standard fitment in much of the US and grew in Europe. This further exacerbated the engine temperature problem and the use of synthetic coolant rather than straight water in the engine cooling system became universal. Its higher boiling point was necessary to handle the higher temperatures.

Over the same period the move to extend the oil drain interval gathered further strength. The driving force for this was primarily the consumer. People had become far more dependent on their cars, and DIY servicing had fallen off rapidly as cars became more complex. Having to put your car in for routine servicing was increasingly viewed as an inconvenience.

The need for longer drain intervals, the more highly stressed engines, and the higher under-bonnet temperatures, all put pressure on the lubricant performance and this was exacerbated by environmental demands. To meet legislation that required the reduced tailpipe emissions to be maintained over extended periods of service, the rate at which engines burnt oil had to be reduced. This was because additives in the oil – particularly those that contained phosphorus – poisoned the exhaust catalyst. As a result, oil top up – which in any case was viewed as an inconvenience by the vast majority of drivers – was reduced, cutting off this avenue of assistance in extending oil drain intervals.

The chemical additive of concern with respect to the poisoning of exhaust catalysts was zinc dithiophosphate (ZDTP), the main antioxidant/anti-wear additive. The presence of phosphorus in the exhaust gases came about because some of the engine oil was burnt from the walls of the cylinders (where it was needed to lubricate the movement of the piston rings over the cylinder bore). The oil and car industries came to a mutual agreement to set a maximum level for phosphorus in engine oils that were sold for use in the new cars. In the US (and much of the developed world), this level was formalised at 0.12 per cent phosphorus in 1989, and reduced further to 0.10 per cent in 1996. However, the oils – with their lower levels of ZDTP – still had to meet all of the engine tests specified in the existing performance classifications, including those for wear of items such as cams and cam followers.

Finally, as if these forces were not enough, car manufacturers also saw a

Which Oil?

potential business advantage if the oil drain interval could be increased to the same as that required for other major service items, such as brakes. They could then potentially cut out the local motor mechanic and gain the full service business. By the end of the period, drain intervals had typically increased to 7500-10,000 miles (12-15,000km).

To meet these requirements and pressures, the performance specifications for engine oils continued to evolve at a rapid pace. To handle the higher temperatures and longer drain intervals, oxidation resistance, detergency and dispersancy increased further, all under the circumstance where the use of a very effective and versatile additive – ZDTP – had to be restricted. Synthetic oils made their first significant appearance in the mass market, as their main strength was performance under elevated temperatures. Most of them also had inherently better viscosity/temperature characteristics (they thinned less at high temperature and thickened less at low temperature), and so they had good fuel economy characteristics, particularly when the engine was cold.

A 'new' problem emerged, initially in Europe: black sludge. While the trend was to higher temperatures, in reality plenty of driving was stop/start or short distance. Even with the better combustion control that electronic fuel-injection gave, combustion by-products found their way into the sump and, with short trips or long periods at idle, engine operation did not reach a sufficient temperature to drive them off. Long drain intervals gave these products time to react with each other and with some of the additive components of the oil. In some circumstances, a black, sludge-like material resulted. Because of the level of dispersant additives present, this sludge was satisfactorily carried around in the hot oil, but it could settle in the colder parts of the engine, such as in the rocker covers. Also, if the drain interval adopted by the driver was too long, the amount of sludge could exceed the oil's capacity to keep it dispersed. Sludge then blocked oil filters and was deposited in the sump and oilways.

The motor industry responded by introducing specific performance tests for an oil's ability to handle black sludge, and these tests became part of the standard performance specifications. As a consequence, the chemistry of some of the additives used in the oils was changed.

At the start of the period the engine oil performance specification was upgraded from SE to SF. In 1987 a further change was made, to SG, and in 1993 another upgrade, to SH. Finally, in 1996, came SJ. Some of these many changes were to handle the demands already outlined, but others were to introduce performance requirements in new areas – areas such as high temperature shear stability, evaporation resistance, filterability, foaming tendency, and flammability (flashpoint), that had not been specified before. Also, compliance with an engine testing code of practice became a requirement with the introduction of SH. Prior to this, an oil could claim to be of a particular performance level on the basis of just a single set of successful tests. Under the code, consistent test outcomes over several tests were needed (or a higher pass level had to be met if just a single set of tests were run). The API 'donut' certification and licensing system for engine oil performance claims was introduced at the same time (1993) in the US.

Historical eras

At the latter end of this period, the US auto industry broke ranks to some extent with the oil industry. It wanted to push the evolution of engine oils at a faster pace, and formed an auto industry-only group (ILSAC). In 1993 it brought out an engine oil specification called GF-1. This was essentially the same as SG, but required that the testing code be used (thus predating SH with this requirement). GF-1 also required a demonstration of fuel economy, and restricted the oils that it listed to those of viscosity grades SAE 0W/XX, 5W/XX and 10W/XX (where XX stands for SAE 20 or 30). This was in an attempt to push consumers to use these lower viscosity products, again for fuel economy. Consumers were resisting this move and the industry – which used these lighter oils in its own testing to demonstrate compliance with mandatory fuel economy legislation – was under pressure to justify the practice.

A somewhat similar move occurred in Europe. The ACEA (the Association of Constructors of European Automobiles) took over the CCMC engine oil specification role and published a new set of specifications. European manufacturers adopted them, but also tended to specify particular 'in-house' tests for oils that were to be used for factory-fill.

With the move to front-wheel drive, transmission designs changed dramatically. They also became more compact and, because of this, were less well cooled. The gear oils had to withstand higher temperatures. However, the need for high load bearing (EP) performance in the final drive lubricant in general was significantly reduced as hypoid (offset) gears largely disappeared.

While there was some reduction in automatic transmissions in the US (with the move to smaller vehicles and the higher cost of gasoline), their numbers increased significantly in Europe, a reflection of the increased congestion on roads. However, in all markets, in the pursuit of fuel economy, these transmissions began to become more complex, with features such as lock up torque converters becoming standard. This led to some modifications to the friction characteristics needed in the lubricants. Five forward gears became the norm in manual transmissions, with the fifth gear being an overdrive to reduce engine speed and save fuel.

Disc brakes displaced drum brakes – completely for the front wheels and to a

---

SG lingers on

SG (introduced in 1989) was the last performance specification that did not require application of the rigorous testing code of practice. To this day you still see some cheaper engine oils on the market that are of claimed SG performance. You should be aware that these claims may not be able to be independently validated. If you use such an oil – and, potentially, they are perfectly suitable for many older cars – I suggest that you buy only reputable brands.

Some problems with SF

It emerged that the SF performance upgrade was something of an aberration. It boosted detergency but not dispersancy, and it may have been a contributor to the 'black sludge' problem. In comparison, dispersancy performance of SG oils was much higher.

Which Oil?

significant extent for the rear – and hub greases moved universally to higher melting point products. Clay greases were largely superseded by so-called complex greases – usually lithium complex – for this application. The need for chassis greases largely disappeared from the motorists perspective as all components involved became sealed-for-life. They had no provision for in-service lubrication.

2.8 Post 1999

This period is relevant to us because it has produced the oils that are on the market today, the oils from which we have to make our selection. I will not cover it in detail other than to say that the big trend has been further fuel economy and, in more recent times, reduced greenhouse gas emissions. For the motor car, greenhouse gas means carbon dioxide (CO_2) and, because the rate of emission of CO_2 is directly related to fuel consumption, it has simply put further pressure on this area.

Worldwide, there has been a significant move by the car manufacturers to using and specifying lower engine oil viscosities. Whereas in 1990 a typical oil viscosity recommendation (in temperate areas) was SAE 15W/40 or 20W/50, today it is SAE 0W/20 or 5W/30. To allow the use of such low viscosities, machining tolerances, surface finishes and metallurgy have all had to improve. Also, to control the loss of these lighter oils at operating temperatures, limits have had to be put on their volatility. This has led to the widespread introduction of full or partly synthetic engine oils.

The oil marketers have recognised that such low viscosity is not necessarily ideal for the engines in earlier cars. There has been some splitting of products into targeted markets, with many oil manufacturers having products for current engines – which they might describe as overhead cam/multivalve – and products for older engines. Also, every major oil brand now has at least one fully synthetic product in its range.

The pressure with respect to emissions other than CO_2 has not diminished either. A further reduction in allowable level of phosphorus in the engine oil (to 0.08 per cent) has occurred – essentially universally in the US with the introduction of the SM engine oil performance classification in 2004, but in a more restricted manner in Europe (the Sequence C or catalyst compatible oils).

Behind the scenes there have been significant changes in the base oils used to formulate all motor car oils – engine and transmission. Since about 2008, there has been increasing use of certain synthetic base stocks, even in oils that are not marketed as synthetics. Conventional, non-synthetic base stocks have also become far more highly refined.

That brings the historical overview to an end. The 100 year history of the motor car has been one of continuous technological advance. This applies as much to its lubricants as it does to the design of the vehicle. Lubricants have become much more sophisticated and much more complex. Unfortunately, at least some of this complexity has to be understood if we are to make safe and appropriate choices of lubricants for our classic cars. The intention of the next two chapters is to provide you with this understanding.

3. Basics of lubrication

When one metal surface moves over another, friction occurs and energy is needed to overcome it. The amount of friction depends on how hard or forcefully the two surfaces are pushed or pressed together. The friction generates heat and, if the load pushing the two surfaces together is high enough, the microscopically-small high points of the two surfaces may momentarily weld together, only for the welds to be immediately broken as the surfaces continue to move relative to each other.

Such movement and friction leads to very high rates of wear of the two surfaces. Also, significant energy is needed to overcome it. However, if the two moving surfaces could be kept separated by a film of liquid (or gas), wear would be eliminated and the only friction involved would be that needed to move the fluid film. This is the simple concept behind lubrication.

If the two surfaces involved can be kept separate by the fluid film under all operating conditions, this is known as full-film lubrication. The conditions that will help to maintain full-film lubrication are a high viscosity (thickness) of the fluid being used as the lubricant, a low force or load pushing the two surfaces together and, perhaps counter-intuitively, a high relative speed between the two surfaces. This latter can be understood if you envisage the situation where the two surfaces are at rest. The fluid will tend to drain away. When the surfaces are moving relative to each other they will entrain or drag the fluid into the area between them.

Given that you cannot readily influence the loads that the moving mechanical components of your vehicle are subject to – they are set by the design – you may think that the obvious thing to do is to use the highest viscosity of lubricating fluid available. However, this ignores the fact that a fluid does have internal friction in its own right. In fact, its viscosity is a direct measure of this friction. Thus, the more viscous the lubricating fluid the greater the frictional losses within it. Also, it is necessary to deliver the lubricant to the critical potential wear areas, and this will be more difficult and energy-consuming if its viscosity is high. In practice, there is an optimum viscosity for every situation.

In summary, for a fluid to act successfully as a lubricant in a given situation it must:

1. Have the appropriate viscosity under the operating conditions.
2. Be physically and chemically stable under the operating conditions.
3. Not react negatively with the surfaces that it is separating.
4. Not react negatively with any of the components of the lubricant delivery system.

To be practical it must also be economic to source and safe to handle.

Further discussion of each of these requirements is given in the sections that follow. However, before going there, it must be noted that, even with perfect design

Which Oil?

and the perfect lubricant, full-film or hydrodynamic lubrication cannot be achieved in all operating circumstances. The most obvious one is at start-up after a period of shutdown. For many components the fluid film will have drained away and, at the instant of starting, the surfaces involved will be in contact. Full film lubrication will only be established once the lubrication system comes into action, and this is almost universally powered by the device itself, whether directly via a pump or indirectly via splash or 'dip,' as in a gearbox or older car engines.

A far less obvious situation where full-film lubrication cannot be established or maintained is between the piston rings and the cylinder bore at the exact top and bottom of the stroke of a conventional reciprocating engine. At the point of reversal of the direction of movement the piston is stationary with respect to the bore or liner, and an oil film cannot be maintained in those locations.

There are also some specific situations where the load or the design in one localised part of a mechanical device is such that a film cannot be maintained by a fluid that has a viscosity suitable for the lubrication of the other components involved. If the viscosity were increased to a level that could handle the localised problem it would be too high for the rest of the components. Examples are tappets, lifters or followers in conventional (sliding) camshaft designs in high-speed engines, and the contacting gear teeth in offset (hypoid) differentials. This situation is further explained later in this chapter (Section 3.5). We return now to the basic requirements that are needed for a fluid to be an adequate lubricant.

As anyone who has slipped on a wet floor can testify, water can be quite a good lubricant. However, if the device being lubricated should operate above 100°C – the boiling point of water – water would be useless, just as it would be if the ambient temperature should fall to 0°C (the water would freeze). As well as this lack of physical stability, water has other problems. These include a relatively low viscosity (limiting its load bearing ability), and a strong tendency to promote the oxidation (rusting) of steel. That is, it tends to react negatively with many of the types of materials used to make the moving surfaces that require lubrication.

In practice, some members of the class of liquids known as oils have proven to be very well-suited to the role of a lubricant. Oils were widely used for purposes such as cooking, lighting, and cosmetics long before the development of mechanical devices. However, they are now so universally used for this 'new' purpose that 'oils' has become the generic name for liquid lubricants.

Up until the development of the mineral or hydrocarbon-based oil industry late in the 18th and early in the 19th century, the oils used as lubricants were the same ones that were used for cooking, etc. They were derived from plants or animals. However, crude mineral oil was found to contain components or fractions that were of an ideal viscosity for lubrication, and that also were chemically and physically stable over a wide range of conditions. These refined or separated fractions from naturally-occurring crude mineral oils have formed the basis of the vast majority of lubricants for machines for the past 80 plus years.

In recent times so-called synthetic oils have begun to play an increasing part. Usually these are also hydrocarbon-based, but, rather than being obtained by

Basics of lubrication

physical separation from natural crude oil, they are manufactured chemically. The raw materials have usually been derived from crude oil, but the finished product is not a component of these oils; it is a synthesised compound. Unfortunately, in the United States there is scope for confusion over what is meant by the description 'synthetic oil.' This is explained in the panel.

Hydrocarbon oils – whether derived from crude oil or via chemical manufacture – make up the major part of the lubricants that we use in motorcars today. When they are used for this role the industry that produces them calls them 'base oils.' The final lubricant that is retailed to the public will possibly contain other components, but most of the product is hydrocarbon oil. Hence these oils form the 'base' of the lubricant.

3.1 Appropriate viscosity under the operating conditions

With this preamble out of the way, we now ask the question: "In practice, what viscosity of a liquid lubricant is found to be ideal to keep moving mechanical surfaces separated?" As has been explained, the answer will depend on the loads and speeds involved. Accordingly, a range of lubricants has been developed with different viscosities. To discuss them we need to have some understanding of the measurement of viscosity.

If you apply a force or a stress to a fluid it will start to move or flow. How fast it moves depends on the property of the fluid that we call its viscosity. If it has a low viscosity it will move relatively quickly, whereas if it has a high viscosity it will move only slowly (in response to the same applied force). The technical definition of viscosity is the ratio of the shear force or stress (that is applied to the fluid) to the rate of relative movement of the fluid (called the rate of strain or shear) that results.

Viscosity = shear stress/rate of strain   (1)

For further details see Figure 3.1 and read the panel.

Definition of synthetic oils

There's no internationally-agreed definition of a synthetic base oil. Historically, the sort of definition that I give in the main text – being produced by chemical synthesis rather than by extraction or refinement of petroleum – has been tacitly accepted, and it still is to a large extent, outside of the US.

However, in the US, the National Advertising Division of the Council of Better Business Bureaus made a ruling, in 1999, that base oils derived by deep hydrocracking of mineral oils could also be called synthetic. As these oils are generally cheaper than 'true' synthetics, great angst has followed, often based more on competitive rivalry than technical analysis. The reality is that the base oil type is not sufficient in itself to determine the performance of a lubricant. All other components must also be selected appropriately.

In my opinion, this is not an argument that need greatly concern you. Rely on reputable brands and standardised industry performance specifications to make your judgement, not on marketing literature. I give more information in later chapters.

Which Oil?

As the panel explains, in the international standardised system of measurement, the units of viscosity are Newton second per m². For simplicity, they are called Pascal seconds, written as Pa s.

A fluid with a viscosity of 1Pa s is fairly viscous. Hence, for practical purposes it's common to use either a unit that is only one-tenth as large (called a Poise, P) or one that is only one thousandth as large (called a centiPoise, cP). That is, 1Pa s = 10P = 1000cP. To give you a feel for these units, at 20°C water has a viscosity of almost exactly 1 centiPoise, while the viscosity of castor oil is about 1000cP (1Pa s), and treacle is 20,000cP (20Pa s).

The viscosity of liquids can be measured in various ways. A simple method is to measure the time that it takes for a ball of known size and weight to fall a certain distance in it. An even simpler method is to measure the time that it takes for a given quantity of the liquid to flow vertically, under gravity, through an orifice or a narrow tube. In this latter type of method the driving force for the flow is the weight of the liquid, and this depends on its density and the force of gravity. The viscosity that is obtained by such gravitational methods is called the kinematic viscosity. The true or absolute viscosity (as defined in equation 1, previous page) can be obtained from it via the relationship:

Absolute viscosity = Kinematic viscosity x Density of the liquid   (2)

The international standardised unit of kinematic viscosity is mm² per second. Again, for practicality, a unit one tenth as large is commonly used – called a Stoke (St) – and also a unit one thousandth as large – the centiStoke (cSt).

These units are international standard measures of viscosity. International agreement to use them was reached only relatively recently, and, prior to this, various countries and regions had their own measures and units. The various units and the relationships between them are discussed in the panel.

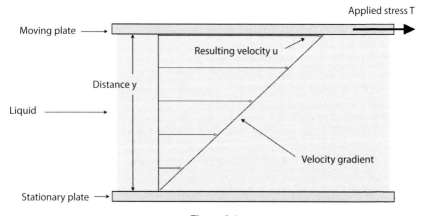

Figure 3.1

Basics of lubrication

Also, the practical use of oils as lubricants grew widely before science fully caught up with it, and the early automotive industry developed its own ways of measuring and specifying viscosity, again often varying region-to-region. Initially, the specification was as simple as 'thick' or 'thin' oil or 'heavy' or 'light.' The first consistent method was to specify the specific gravity range that the oil should fall into. The more dense the oil the more viscous it was. However, this was not precise

Definition of viscosity

In Figure 3.1, the moving (top) plate applies a shear stress T to the fluid which moves at velocity u, where it is in contact with the plate, but remains at zero velocity against the stationary (bottom) plate. If the distance between the plates is y then the resulting rate of strain of the liquid is u/y. The definition of viscosity is (see Equation 1, p23):

Viscosity = shear stress/rate of strain   (3)
= T y/u

The equation shows that, for a given applied stress T, the resulting velocity u is faster when the viscosity is lower.

The units of the applied stress are force per unit of area of the fluid in contact with the plate, or N/m^2 (where N is Newtons). The unit of distance is m and of velocity is m/sec. Putting these units into Equation 3 shows that the units for viscosity are N sec per m^2.

Early methods of viscosity measurement

In the US, the Saybolt viscometer was the instrument of choice in the early years of the motor car. It had been developed by Mr GM Saybolt around 1880. At a lecture given by Mr Boverton Redwood in 1886 in England, where he demonstrated his own new viscometer, he also demonstrated the Saybolt one. International exchange of ideas was already well established.

Whereas Saybolt's device measured the time for the liquid to flow through a vertical tube, Redwood's used an orifice.

In Germany, Engler developed a similar device, also based on an orifice, but he reported the flow time as the ratio of the time taken for the liquid itself to the time taken by water at the same temperature. This had the advantage of not requiring highly reproducible dimensions of the orifice, instrument to instrument. The results are expressed as Degrees Engler, the ratio of the times. Thus water has a viscosity of 1 deg Engler.

Some approximate relationships between the various units of measurement are:

10cSt = 1.83 Degrees Engler = 59 Saybolt Universal Seconds = 52 Redwood No 1 Standard Seconds

However, the relationships vary with the actual magnitude of the viscosity being converted.

Which Oil?

because oils derived from different crude oils (or from plants or animals), could have quite different viscosities and yet have the same density or specific gravity. As time passed and performance demands increased, methods and terminology were formalised, and eventually standardised on actual measurements of viscosity. While some of the modern terminology relates back to old methods, it is now based on standardised international definitions.

The most widely used system of definition of the viscosity of oils for motorcars is that of the American Society of Automotive Engineers (SAE). It covers two families of lubricating products – oils for engines and oils for gears. The SAE system assigns

SAE viscosity grade	Low temperatures		High temperatures		
	Cranking viscosity (cP max at °C)	Pumping viscosity (cP max at °C)	Viscosity (cSt min) at 100°C	Viscosity (cSt max) at 100°C	High shear viscosity (cP min at 150°C)
0W	6200 at -35	60,000 at -40	3.8	-	-
5W	6600 at -30	60,000 at -35	3.8	-	-
10W	7000 at -25	60,000 at -30	4.1	-	-
15W	7000 at -20	60,000 at -25	5.6	-	-
20W	9500 at -15	60,000 at -20	5.6	-	-
25W	13,000 at -10	60,000 at -15	9.3	-	-
20	-	-	5.6	<9.3	2.6
30	-	-	9.3	<12.5	2.9
40	-	-	12.5	<16.3	2.9*
40	-	-	12.5	<16.3	3.7**
50	-	-	16.3	<21.9	3.7
60	-	-	21.9	<26.1	3.7

*0W/40, 5W/40 and 10W/40 grades only
** all other 40 grades

Table 3.1 SAE J300 Viscosity grades for engine oils.

Early specification of viscosity

The owner's manual for a 1911 Model T Ford states:

"We recommend only light high-grade gas engine oil for use in the Model T motor. A light grade of oil is preferred as it will naturally reach the bearing surfaces with greater ease, and, consequently, less heat will develop on account of friction. The oil should, however, have sufficient body so that the pressure between the two bearing surfaces will not force the oil out and allow the metal to come in actual contact. Heavy and inferior oils have a tendency to carbonize quickly, also gum up the piston rings and valve stems."

Basics of lubrication

a simple number to a particular viscosity range. The numbers for engine oils, and the viscosities to which they correspond, are listed in Table 3.1.

You'll note that the viscosities that define which particular SAE grade an oil

---

A brief history of the SAE viscosity system

It may be of interest to know where the SAE engine oil viscosity numbers came from. Even today, the US has not adopted the metric or ISO system of units, and this was certainly the case in 1911 when the SAE viscosity classification system was born. The first system was in some ways more comprehensive than the one we know today. As well as viscosity at 100°F and 210°F, it specified properties such as specific gravity, flash and fire tests, and carbon residue – a reflection of the uncertain quality of the engine oils at that time. It did not specify low temperature properties, other than to list two grades with defined Pour Points.

If the US has been slow to metricate, it was quick to adopt reproducible scientific techniques of measurement. In 1911, the measurement of the viscosity of oils was standardised on an instrument developed by GM Saybolt. The time for 60ml of an oil to pass under gravity through a standard tube was measured, with the result being called Saybolt Universal Seconds (SUS or SSU).

In 1911, ten different viscosity grades were specified in the new SAE system. For the lower viscosity ones, the grade was designated by the first two digits of the average SSU viscosity at 100°F. For example, the grade defined as having a viscosity at 100°F of between 180 and 220SSU was designated Grade 20 because its average viscosity was 200SSU. The grade number was essentially the average viscosity at 100°F in SSU, divided by ten. A different base was used for the numbering system for heavier grades. It related to the viscosity at 210°F.

The system was unnecessarily complex. In 1926, it was revised to a basis that we would recognise today. The old grade numbers – 10, 20, 30, etc – were retained, but they no longer related directly to viscosity. The use of SSU as the measure of viscosity persisted through to 1976. Given that the Saybolt method of measurement uses gravity, its result is related to the kinematic viscosity of the liquid.

Winter or W grades

The first use of the W viscosity grades and notation occurred in 1933 (although there was a delay in their commercial introduction). Two new grades were defined; 10W and 20W. They were specified by a maximum and minimum viscosity, in SSU units, at 0°F. The values were obtained by extrapolation from higher temperatures. Thus the 10W specification was simply the expected viscosity of an SAE 10 oil at 0°F and 20W was based on the expected viscosity of an SAE 20 oil at 0°F.

As base oil refining has advanced over the years since 1933, this simple direct relationship has been lost, although the basic concept remains. As well as meeting the specified low temperature requirement, a 20W oil must also have the viscosity at 100°C of an SAE 20 oil. However, an SAE 20 oil will not necessarily meet the 20W specification. It may be too viscous unless it has been specially refined or is an appropriate synthetic.

Which Oil?

lubricant belongs to are measured at 100 and 150°C for the grades that do not have a W in their designation, and at very low temperatures (as well as at 100°C) for those that have a W. The grades with W in their description are often called Winter grades. It's possible for a particular oil to have viscosity characteristics that fall into both a W grade and a non-W grade. Such oils are called multigrade oils, and I will discuss them in greater detail in Section 4.1.

3.1.1 The effect of temperature on viscosity

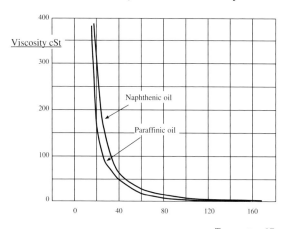

Figure 3.2

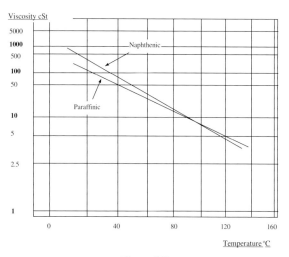

Figure 3.3

The viscosity of all liquids, including oils, changes with temperature. For oils, the change is very marked. For example, a typical mineral oil that has a viscosity of 120 cSt at 20°C would have a viscosity of only 10 cSt at 120°C, a more than ten-fold decrease. In many parts of the world, mechanical devices that require lubrication could easily see such a range of operating temperature. When starting up after a period of non-use, the moving surfaces could be at 20°C or lower, whereas, after a period of operation, they could be at 120°C or higher. The lubricant must keep the surfaces separated and protected at both extremes of the operating temperature range.

Because the change in viscosity is so large it can be difficult to compare different oils by simply plotting their viscosities against temperature. An example is given in Figure 3.2 (the terms

Basics of lubrication

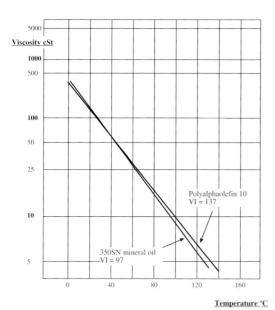

Figure 3.4

naphthenic and paraffinic are defined in the panel in Section 2.1). To make comparisons easier, it is common to take the logarithm (twice, in fact) of the viscosity and the logarithm of the temperature and plot these values instead. This tactic changes the curved relationship to a linear one. The outcome for the curves that were plotted in Figure 3.2 is shown in Figure 3.3.

It's important to realise that this is simply a mathematical manipulation. The large change in viscosity with temperature remains and is real. In practice, this means that the oil chosen will have to be one with adequate viscosity at the higher temperature, and its performance at the low temperature may then be inefficient.

Given this, it's no surprise that attempts have been made to formulate oils that change less in viscosity with temperature. The most common approach – which has been used widely since the 1960s – is to add a substance to the base oil that is called a viscosity index improver (VII). Just what these viscosity index improver substances are and how they work are both covered in the next chapter (Section 4.1).

Viscosity index or VI is a measure of how much an oil changes in viscosity between 40°C and 100°C. It's explained further in the panel. Slightly simplistically, the higher the viscosity index of an oil, the less is its change in viscosity with temperature.

An alternative approach to this problem of thinning with increased temperature is to synthesise base oils whose viscosity inherently changes less with temperature than is the case with oils derived from natural mineral oil. This then avoids or reduces the need to add another substance. Commonly used synthetic base oils are ones based on polyalphaolefins (PAOs). A typical PAO has a viscosity index (VI) of greater than 130, while a typical paraffinic mineral oil has a VI in the range 95 to 100. Figure 3.4 shows a comparison. However, not all synthetic base oils have inherently-high VIs.

Just as oils thin with increasing temperature, they also thicken as the temperature

Which Oil?

is lowered. Also, with mineral oil-based lubricants there is another complication. These oils generally contain waxes, and at low temperature the wax will crystallise and come out of solution as a solid. In other words, the physical form of the lubricant is not stable at low temperatures.

For an engine there are three routes by which such a cold, thick oil can cause problems. The first is the obvious one that the viscosity may be so high that the mechanical lubricant circulation system cannot pump it to the critical parts at a sufficient rate. The second route can be that the oil simply cannot flow to the pick up point of the pump at a sufficient rate and the pump starves or cavitates. The third arises in the situation where wax has precipitated out of the oil. It may block the oil filter or, if there's a strainer on the inlet to the oil pump, it may block this. Such blockage can give partial or complete lubricant starvation.

If you look at Table 3.1 you'll see that the W oils must meet two different cold temperature viscosity specifications. These are designed to cover the first two possible routes to a performance problem. The pumping measurement (carried out by using an instrument called a Mini Rotary Viscometer (MRV)) relates to the pumping or circulation viscosity, while the cranking measurement (using a Cold Crank Simulator (CCS)) relates to the flow of the oil under just its own head, which is

---

Viscosity index

The viscosity index (VI) of an oil is a somewhat arbitrary measure or indication of the relative tendency of its viscosity to change with temperature. In Figures 3.2 and 3.3, the viscosity of a paraffinic and a naphthenic oil are plotted against temperature. (For an explanation of the terms naphthenic and paraffinic, see the panel in Section 2.1).

From Figure 3.3 in particular, you will see that the viscosity of the paraffinic oil changes less than the naphthenic. It would be useful to have a simple parameter that shows this relative tendency. The parameter used in the oil industry is the viscosity index.

Originally, the industry took two standard oils – one highly naphthenic and one highly paraffinic – and measured their viscosities at 40°C and 100°C*. It declared the particular naphthenic oil to have a viscosity index of zero (0) and the paraffinic oil to have a viscosity index of 100. To find the viscosity index of any other oil, its viscosity at 40 and 100°C is measured, and its relative ranking, between the two standards, is estimated. To assist with this, tables and formulae are available. For the two oils plotted in Figures 3.2 and 3.3, the calculated Viscosity Indices are 27 for the naphthenic and 95 for the paraffinic.

As base oil refining progressed – and also with the introduction of synthetics and VI Improvers – oils were produced that had a viscosity change that was less than that of the oil that had been used historically to set the 100 reference. These oils had VIs greater than 100. To avoid the need for extrapolation, for oils with a VI above 100 a relatively simple formula has been devised. Provision of the formulae or the tables is beyond the scope of the book. For your purposes, the important fact to remember is that, in general, the higher the viscosity index of an oil, the less it will thin as temperature increases.

* Actually, in the original definition, the reference viscosities were 100°F and 212°F, but they have since been standardised at 40°C and 100°C.

Basics of lubrication

the situation involved with the flow of the oil in the sump to the pump inlet. Neither of these two tests – MRV and CCS – involves the flow of oil under gravity. Hence, the viscosity units in the specification are centiPoise.

The third potential problem – the wax precipitation tendency – was specified historically by a property called the Pour Point. This was replaced briefly by the Stable Pour Point but, since the late 1990s has been measured by a property called the Gellation Index. Given that wax precipitation is a form of physical instability of the lubricant, it is dealt with more fully in Section 3.2.2 following.

3.1.2 Relation of viscosity to fuel economy

I have mentioned that energy is required to overcome the internal friction of the lubricant, and that the magnitude of this internal friction is related directly to the viscosity of the oil. Since the first oil crisis in 1973 there has been pressure to improve the fuel economy of motor vehicles. This pressure has been ramped up since 2000 as the need to reduce carbon dioxide (CO_2) emissions has also been recognised. For motor vehicles, the amount of CO_2 emitted is directly dependent on the vehicle's fuel consumption.

Reducing the viscosity of the lubricating oil used in the engine is one way of reducing fuel consumption and, compared with engineering changes, it's a very cheap option. Hence, there has been a very marked tendency over the past five to ten years for vehicle manufacturers to recommend much lower viscosity engine oils than they have historically. In order to facilitate this the manufacturing tolerances, metallurgy, and surface finishes in the engines have had to be improved significantly. Also, because components still have to be protected at high temperatures, the engine manufacturers have also required less thinning of the oils with temperature – that is, they have specified higher VIs.

A potential problem with low viscosity or light oils is that they are more volatile than heavier ones. Oil is gradually lost in operation through evaporation, particularly from the cylinder walls. This loss not only requires replacement but also contributes to emissions. Accordingly, volatility specifications have been introduced for engine oils in recent years. This has increased the use of synthetic or part-synthetic base oils as these usually are less volatile than those derived from mineral oils.

In the drive for fuel economy, much greater attention has also been paid to the aerodynamics of motor vehicles. To reduce drag, bodies have become much more enclosing of the mechanical and suspension parts. As a consequence, items such as transmissions are less exposed to airflow and can run much hotter, to the detriment of their lubrication.

Finally, some oils have been formulated to specifically reduce the friction in engines. This is dealt with in Chapter 4 (Section 4.6).

3.2 Stability of the lubricant

3.2.1 Chemical stability

A lubricant must remain physically and chemically stable under the operating conditions for which it was designed if it is to retain its desired characteristics. One

Which Oil?

of the main reasons that mineral oils replaced vegetable oils as lubricants was their much greater chemical stability at high temperatures. However, this great chemical stability is also a problem when it's time to dispose of the used oil. It's a little ironic that vegetable-based oils are being revisited as lubricants because they are biodegradable.

Mineral oils are derived from a crude oil that has been buried in the earth for many millions of years. Anything that has survived for so long is, almost by definition, going to be chemically stable. However, there's not very much oxygen below the earth's surface, and it is oxidation – reaction with oxygen – that is a mineral oil's Achilles heel. However, this reaction does not come into play until quite high temperatures. At ambient temperature, a mineral oil will last for many dozens of years if stored in an appropriate container. Most synthetic oils – being built up from components that have been derived from crude oil – are also vulnerable to oxidation at high temperatures.

Oxidation is a reaction that increases in rate very markedly with increasing temperature. For a modern mineral base oil it becomes quite noticeable at about 120°C and increases rapidly above that temperature. You might ask: "Why would oxidation be a concern if it did happen?" The answer is that many of the substances that are formed when oxygen reacts with the molecules that make up a mineral oil are far more viscous and 'sticky' than the original oil. As it oxidises, the oil thickens and tends to coat and adhere to surfaces. Also, some of the substances formed by reaction of mineral oils with oxygen are acidic, especially in the presence of moisture, and they tend to attack many metals. The overall result from ongoing oxidation is that the oil becomes less and less appropriate as a lubricant.

There are specific chemical tests to detect oxidation of a mineral oil-based lubricant. As a user, what you will observe – if the oil in your engine is oxidised – is a significant darkening (to essentially black), an obvious thickening and a distinctive burnt smell. For a gear or differential oil, there may also be a strong, very unpleasant, rotten egg gas type of odour. This is from the oxidation of the additives, not the base oil.

Given the ultimate vulnerability of mineral oil-based lubricants to oxidation, it's logical that the oil industry has worked hard on this problem. All reputable hydrocarbon-based lubricants sold today, including synthetic oils, contain added chemicals whose role is to prevent or greatly slow down oxidation. They are called, logically enough, antioxidants, and they are discussed further in Chapter 4.

Even when there is no oxygen present an oil may change chemically if it is heated. This is known as thermal instability. In comparison with many other substances, mineral oils are very thermally stable and the level of stability can be enhanced further by the refining processes that are used to produce them from crude oil. As has been mentioned, superior thermal stability was one of the reasons mineral oils displaced animal and vegetable oils as lubricants. However, at the temperatures encountered in an engine, mineral oils will, over time, undergo some thermal change. Synthetic oils are, in general, significantly more stable in this respect and this was a prime reason for their initial development. Synthetic

Basics of lubrication

engine oils were introduced first in aero engine applications where temperatures are higher than the automotive equivalent. Oddly, they often also have very good low temperature properties and the other area of early adoption was for oils for use in very low temperature regions of the world, such as the Arctic.

As well as chemical changes of the lubricant itself, changes can occur in service via contamination. All hydrocarbon-based fuels contain some sulphur. Diesel fuel contains far more sulphur than gasoline, but gasoline does contain some. When sulphur is burnt it forms an oxide of sulphur. In the presence of water – which is also produced when a hydrocarbon fuel is burnt – the sulphur oxide forms an acid (sulphuric or sulphurous acid). These acids will corrode metals, particularly soft metals such as those used in shell bearings, leading to greatly increased rates of wear.

All modern engine oils contain chemicals that will react with acids and neutralise them, eliminating their corrosive property. These chemicals are called bases, and the capacity of an oil to neutralise acids is given by a property called its Total Base Number or TBN.

3.2.2 Physical stability

As mentioned in Section 3.1.1, an example of potential physical instability of mineral-based oils is precipitation of wax at low temperature. For those readers who live in temperate climes this is not a concern. The waxes in mineral oils – which are paraffin waxes – do not come into play until ambient temperatures drop below $0°C$ ($32°F$). However, if you do live in a region where such temperatures occur in winter, and you wish to drive your vehicle during that period (or even just run the engine), the low temperature properties of all of the lubricants used in it, but particularly in the engine, are critically important.

The temperature at and below which wax will render a lubricant almost certainly unsuitable for use is defined as its Stable Pour Point. It is measured by a very simple but long-winded test wherein a sample is stored in a standard container for an extended period of several days while the temperature is slowly cycled in the range 0 to $-23°C$, and then an attempt is made to pour it. The pouring attempt is commenced at $-12°C$ in the final cooling cycle, and is repeated at decreasing temperatures, each $3°C$ lower than the previous, until the oil fails to pour. The reported Stable Pour Point is the lowest test temperature at which the lubricant still does pour.

Given the extended time necessary to conduct the test, it has been replaced by a measurement called the Gellation Index. The test measures the viscosity of a sample as it is cooled at a constant rate of $1°C$ per hour. The gelation point is defined as the temperature at which the sample reaches a viscosity of 30,000cP and the gelation index is defined as the largest rate of change of viscosity increase over the range from $-5°C$ to the lowest test temperature.

The base oils used in all modern oils are dewaxed during the refining process, but the finished lubricants also contain substances that are added by its manufacturer to depress or lower the inherent Pour Point of the base oil. These substances, which

Which Oil?

are called Pour Point Depressants, work by interfering with the growth of the wax crystals as they come out of solution. The crystals remain small and do not interlace with each other. Hence they do not interfere as much with the flow of the oil.

It is generally the case that an oil may still be pumped at a temperature a little below its Pour Point, particularly if it contains a Pour Point depressant. Mechanical force can move the oil where gravity cannot. However, it would be very risky to use a lubricant in service at an ambient temperature below its Stable Pour Point. Fortunately, physical instability through crystallisation of wax is a completely reversible process. Once the temperature rises above the crystallisation temperature the wax will redissolve in the base oil. Hence, from the lubricating perspective, it is not a particular concern if a vehicle sits, unused, at a temperature below the Pour Point of its engine oil as long as it is not started until the temperature rises again.

Synthetic oils usually have much lower inherent Pour Points than those based on mineral oils. Oils for Arctic or Antarctic use are almost invariably synthetic-based.

While wax is a concern only in very cold climates, there is a form of physical change that happens to all lubricants used in internal combustion engines. Some products of the combustion of the fuel end up in the oil in the sump. This cannot be avoided. Soot is the main example, and it's inevitable that some of it will end up in the lubricant. If it is allowed to build up it will cause abrasive wear in its own right but will also thicken the lubricant, increasing its viscosity. Handling of soot in the formulation of a lubricant is dealt with in the next section.

During cold starting, particularly in carburetted cars, some unburnt fuel may leak past the piston rings and also finish up in the sump. This will dilute the oil and lower its viscosity. If the fuel contains a high percentage of ethanol, other problems can occur (see the panel).

Direct foreign contamination can be another cause of physical change. Solid contaminants, such as sand and dirt, may enter the engine via the air intake if filtering of the combustion air stream is inadequate.

A different form of physical instability in lubricants that contain viscosity index improvers is the loss of viscosity caused by physical shearing of the polymers that make up the VI Improver. This is covered in Section 4.1.1.

Finally, oils are usually being churned and splashed when in use and can entrain air. If the air built up in the oil, then the oil's ability to keep surfaces apart would be greatly reduced and wear would occur. Since entrainment of air cannot be avoided, it's important that the oil has a good ability to shed it. To assist with this, chemicals called antifoam agents are added. They are usually silicone-based (although some other compounds are used) and only very small amounts are needed.

3.3 Protection of surfaces

One of the important jobs of the lubricant in relation to the internal surfaces of a mechanical device – in addition to keeping moving surfaces apart – is to keep them clean. This is a particularly challenging task in an internal combustion engine, where substances formed from combustion of the fuel can deposit on surfaces and interfere with the movement of mechanical components or with heat

Basics of lubrication

transfer (cooling). Also, as mentioned in Section 3.2.1 under Chemical stability, the breakdown of the lubricant itself from oxidation produces substances that are especially adhesive to surfaces.

In the 1960s special cleaning chemicals that could work at the high temperatures experienced in internal combustion engines became commercially available. They are called detergents and, today, all engine oils contain them. Detergents are not usually added to gear or diff oils because substances that have a tendency to deposit on surfaces are generally not formed in the absence of combustion, provided that the oil itself does not oxidise.

As I mentioned in the previous section, an unavoidable contaminant that arises from combustion – particularly in diesel engines but also to a lesser extent in gasoline engines – is soot. Soot is carbon, which can form if hydrocarbon fuel (or hydrocarbon lubricant) is burnt with insufficient air (oxygen) present.

Physical removal of soot and other solids by filtration is one approach. However, for this to work, the soot particles – which, when they initially form, are extremely small – must agglomerate to a size that the filter can trap. However, as they agglomerate, there is obviously an increased danger of deposition on surfaces, and also of abrasion of moving surfaces as the particles circulate with the lubricant. Accordingly, from the 1960s on, substances called dispersants have also been widely used in engine oils. These keep soot and other organic dirt particles finely dispersed and hence suspended within the oil. If there is a filter in the system and dispersants are present in the oil, it will not remove a significant amount of soot. The soot will be too fine. The filter's main job in such a situation is to remove other solid contaminants. Note that this is the reason that modern oils can become quite dark in colour during service: soot remains in them, albeit as very fine particles.

Another unavoidable consequence of the combustion of fuel is the production of water. Water and some of the other substances formed in combustion can attack metal surfaces, via either rusting or other forms of corrosion. Accordingly, all modern oils contain antirust and anticorrosion additives. Also, many of the metals used in a typical engine or gearbox – iron and copper in particular – catalyse or speed up

---

Potential problems with ethanol in the fuel

Although the problem of phase separation of gasoline fuels containing ethanol and other oxygenates when they become contaminated with water is reasonably well publicised, this is not the case with respect to possible problems with engine oils. Under extreme circumstances, an engine oil that becomes contaminated in use with an oxygenated fuel and some water may also separate into two phases – that is, a phase consisting mainly of water and a phase consisting mainly of oil.

While the engine is not in use, the water phase will sink to the bottom of the sump on standing, with potentially disastrous consequences on restarting the engine. Unfortunately, the lube oil pick-up is invariably also positioned at or near the bottom of the sump, and so initial lubrication of the engine will be with water. This is another reason to use highly-oxygenated fuels with caution.

Which Oil?

oxidation of the oil. Substances to stop this action are also added to modern oils. They are called metal passivators or deactivators.

I am conscious that it is very unlikely that you have been able to digest, let alone remember, all of this information on the various additives that are used in formulating a modern engine oil. To hopefully assist you, I have summarised their role or function in Table 3.2. Potentially of more interest is Table 3.3. This shows the approximate dates at which the use of each particular additive type became widespread in the engine oils sold to the public. Knowing the date when your vehicle was manufactured, you can use Table 3.3 to get some idea of what chemicals were present in the engine oil that it was probably designed for, and also see what new performance additives have come into play since.

3.4 Protection of the lubricant delivery system

In many situations a liquid lubricant would leak out from between the surfaces that it is lubricating if it were not constrained. The constraint is often achieved by the use of seals. Also, there is a general need for access to the internal parts of the mechanical device for assembly and repair. This leads to the need for openings and removable covers, which must also be sealed. This is usually done with gaskets.

There are many designs of seals and gaskets. Some of them are mechanical, but in vehicles they are more commonly made from materials such as rubber (historically natural but today invariably synthetic), cork and felt. The lubricant must be compatible with all of these materials. For example, with seals, it must not cause them to change their dimensions excessively or react with them chemically. If there is to be any change, it should be a slight swelling rather than a shrinkage. To ensure this, modern oils contain a small amount of seal swell additive. However, because

Additive	Function
Antifoam	Prevent persistent foam forming in the oil
Antioxidant	Reduce rate of oxidation of the oil
Antiwear	Reduce friction and wear under partial film lubrication conditions
Corrosion/rust inhibitor	Prevent corrosion and rust of metal parts that are in contact with the oil
Detergent	Keep surfaces clean
Dispersant	Keep solid particles dispersed in the oil
Extreme pressure agent	Reduce friction and wear under boundary lubrication conditions
Friction modifier	Alter friction characteristics of the oil
Metal deactivator	Reduce catalytic effect of metals on oxidation rate of the oil
Pour point depressant	Enable flow of the oil at lower temperatures
Seal swell agent	Swell rubber-type seals in contact with the oil
Viscosity index improver	Reduce the rate of viscosity change of the oil with temperature

Table 3.2 Function of various additives.

Basics of lubrication

modern engine seals are made from synthetic rubber, this is the type of seal that the additives are designed for.

3.5 Imperfect lubrication

Earlier in this chapter I mentioned that there are some specific situations where the load or the design in one localised part of a mechanical device is such that a lubricating or separating film simply cannot be maintained – between the moving surfaces involved – by a fluid that has a viscosity that is also suitable for the lubrication of the other components in the device. If the viscosity were increased to a level that could handle the localised problem it would be too high for the rest of the components.

Hopefully, you'll also recall that the conditions that will help to maintain full-film lubrication are a high viscosity of the lubricating fluid, a low force or load pushing the two surfaces together, and a high relative speed between the two surfaces. These three parameters can be combined into a single factor which allows the lubrication situation to be explained more clearly and succinctly, as in Figure 3.5. Such a plot is known as a Stribeck curve.

Perhaps the best way to understand what the curve shows us is to envisage a lubrication system where the lubricant viscosity and the load are fixed, and we vary the speed of the two surfaces, starting at rest and gradually increasing their relative

> **Detergents vs dispersants**
>
> In the main text I have described detergents and dispersants as having different and separate functions – detergents clean surfaces and dispersants keep solid organic particles separated from each other and suspended in the bulk oil. While this is broadly true, the reality is that the products are not quite so separate in their characteristics and functions. Detergents, by their manner of working, do have some dispersancy properties, and many of the substances used in oils have both characteristics. Dispersants have less inherent detergency properties but they do have some. A chemical difference is that detergents contain a metal – typically calcium, magnesium or sodium – whereas dispersants do not.

Year	Anti-foam	Pour point depressants	Anti-oxidants	Zinc/phosphorus antiwear	Anti-corrosion	Detergents	Dispersants	Viscosity index improvers
1925	✓	✓						
1930	✓	✓	✓					
1935	✓	✓	✓	✓	✓			
1960	✓	✓	✓	✓	✓	✓		✓
1965	✓	✓	✓	✓	✓	✓	✓	✓

Table 3.3 Approximate year of introduction of various additives to commercial engines.

Which Oil?

speed. This means that we will gradually move along the curve, starting at the left where, because the speed is zero, the Stribeck parameter is also zero, and then moving to the right as the speed, and hence the parameter, increases.

At rest, the oil film drains away, and so, at initial movement, the surfaces are in contact and the friction is highest. At very low speeds lubricant still does not penetrate between the surfaces, but the friction drops because, as the speed increases, the surfaces interact less. They tend to 'skim' over each other. However, friction is still relatively very high, and wear would be extreme. This region of operation is called the boundary region.

As speed increases further, some lubricant is entrained into the still very small gap between the surfaces. It reduces the friction but does not produce complete separation of the surfaces. This region is called the mixed lubrication or mixed friction region.

With further increase in relative speed, more lubricant is entrained, friction drops, and eventually the situation is reached where the surfaces are fully separated. At this point the friction is at its least. It is the end of the mixed region and the beginning of the full-film or liquid friction region. Further speed increase sees an increase in friction. This is due to the internal friction or viscosity of the lubricant. The faster it has to be moved the greater the energy needed and hence the greater the perceived friction. However, it's much less than was the case in the boundary zone.

Ideally, you would like to always operate a mechanical device in the full fluid region. This is not possible at start-up (nor at shut down). Also, not all of the areas in a mechanical device where surfaces move relative to each other are equally loaded.

If you look at Figure 3.5 you'll see that, for a given speed and viscosity, the value of the Stribeck parameter decreases as the load increases, and so you move to the left on the curve. If the load is sufficient, you move out of the liquid friction (full-film) region and into the mixed region.

An example of such an effect exists between the cams and the sliding cam followers in a typical overhead valve engine. While the speeds are high and the temperatures reasonable (so that the viscosity of the lubricant is

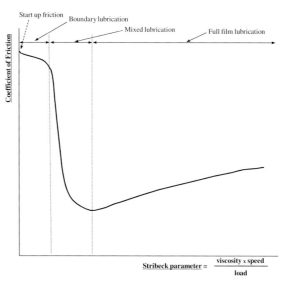

Figure 3.5

Basics of lubrication

maintained), the loads are such that the surfaces involved are in a mixed lubrication regime. At first sight the loads may not appear to be very high but they are being applied to quite small areas. This is one of the reasons that high-performance engines have adopted roller cam followers in the past. Roller followers also reduce friction and improve fuel consumption and so, today, many mass production cars also have them.

There's another common situation where full-film lubrication cannot be maintained. It is due to a geometrical problem and exists in almost every modern rear-wheel drive motor car. It occurs in hypoid differentials. In a hypoid gear set the pinion gear is offset from the centre line of the bevel gear (see Figure 3.6b). In order to accommodate this, the spiral angle of the pinion gear teeth is not the same as those on the bevel gear. As a result, there is a sliding motion between the teeth on the pinion and those on the bevel during the period in which they are in contact. The amount of sliding depends on the amount of offset of the pinion shaft.

This sliding motion tends to wipe the lubricant film from the surfaces of the gear teeth during their contact period and the boundary lubrication region is entered. The problem is compounded if contact loads are high (which they often are in a final drive).

This problem does not occur if there is no offset of the pinion shaft (see Figure 3.6a). The spiral angles of the two gears can then be the same, and there's no sliding action. This type of gearing is called spiral bevel and it exists in the differentials of some older cars, typically pre-World War II. The offset of the pinion shaft was introduced primarily to allow cars to sit lower to the ground. It also increases the tooth contact area.

Faced with these problems, all modern engine and gear oils contain additives to reduce the wear that would otherwise occur under these unavoidable partial lubrication conditions. They are discussed in Sections 4.3 and 4.4 following.

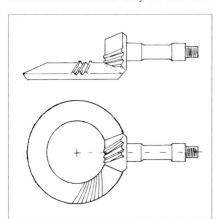

Figure 3.6a Spiral bevel gear set.

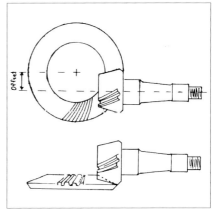

Figure 3.6b Hypoid gear set.

4. The roles of lubricant components

As I explained in the previous chapter, there are situations where the load or the design in one localised part of a mechanical device is such, that a lubricating or separating film simply cannot be maintained between the moving surfaces involved by a fluid that has a viscosity suitable for the lubrication of the other components in the device. Examples mentioned were sliding camshaft followers and the gear teeth in hypoid differentials. Also mentioned were the great desirability of reducing the amount by which a lubricating oil thins as it gets hotter, and the susceptibility of a base mineral or hydrocarbon oil to oxidation at these higher temperatures.

To deal or help with these and other problems, various components have been developed to add to the base oil. These are components which the formulator of the lubricant adds to give the performance level claimed for the product, and they are an essential and integral part of the product. I am not talking here about additives sold in the aftermarket.

4.1 Reduction of oil thinning with temperature
4.1.1 Viscosity index improvers

As was explained in the panel in Section 3.1.1, viscosity index is a comparative measure of the extent to which the viscosity of an oil changes with temperature. The substances added to base oil to reduce its change of viscosity with temperature are called viscosity index improvers (VIIs). These synthetic polymers are chemicals whose molecules are relatively very large, the molecules being made (that is, synthesised) by linking together a large number of identical smaller molecules.

For some polymer molecules, their physical form when cold is a coiled up 'chain.' As the temperature is increased the chain tends to uncoil and straighten out. If a quantity of such a polymer is mixed into a base oil, but does not dissolve fully in it, then at low temperatures its molecules will be small and not interfere much with the flow of the surrounding base oil. They will not increase the viscosity of the base oil a great deal. However at higher temperatures, when the molecules tend to uncoil, they will interfere with flow and hence increase the viscosity of the base oil relatively more than they do at low temperatures. This increasing thickening contribution as temperature rises will compensate, at least in part, for the natural thinning of the

> **More regarding mode of action of VI improvers**
> Although a mechanism of action based on coiling and uncoiling polymer molecules is relatively easy to describe and imagine, the reality of the situation is more complex than this. Change of solubility of the polymer in the base oil with temperature also has an important part to play. However, there is little to gain from a more detailed explanation, and the one based solely on change of shape is easy to visualise and remember.

The roles of lubricant components

Base oil groups

Although modern engine oils contain many additives, the base oil remains the key component. A base oil may be produced by refining mineral oil or by chemical synthesis. Also, different degrees of refining are possible.

Given these different possible production routes, base oils are today classified into five groups (six in Europe). The basis of the classification is shown in the table.

Group	Viscosity index	Saturates		Sulphur	Type
I	80-120	<90 per cent	and/or	>0.03 per cent	Mineral oil
II	80-120	>90 per cent	and	<0.03 per cent	Mineral oil
III	>120	>90 per cent	and	<0.03 per cent	Mineral oil
IV					Polyalphaolefin (PAO)
V					All other synthetics, except PIO
VI			Europe only		Polyinternal olefins (PIO)

Mineral oils still make up by far the majority of base oils used. However, in the past few years, there has been a dramatic move away from Group I to Group II or Group III. This has been driven as much by changes in the economics of the oil refineries that produce the base oils as it has by the need to meet higher performance specifications.

base oil. The outcome is less reduction or change of the viscosity of the base oil/polymer mixture with increase in temperature when compared with the case where no polymer is present.

The effect can be shown most readily on a plot of the double logarithm or linear type that was described in Section 3.1.1. An example is shown in Figure 4.1. Adding VI Improver polymer to a mineral oil increases its viscosity at all temperatures, but the

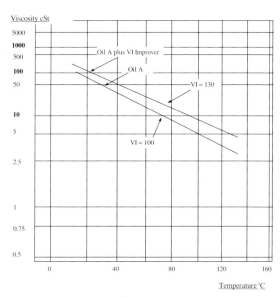

Figure 4.1

Which Oil?

relative thickening is greater at high temperatures, giving a reduction in the slope of the line and a higher VI.

Given that the thickening effect is caused by the polymer chains interfering with the flow of the base oil, it's probably obvious that such an effect will be greatest if the chains are orientated at 90 degrees to the flow direction of the base oil, and will be the least if they are aligned with the direction of flow. When mixed with the base oil – and with the base oil at rest – the polymer chains will take up random orientations and so their thickening effect will be somewhere between the two extremes (90 degrees to the flow and parallel with the flow). However, once the oil is in motion, the internal drag will tend to orientate the chains in the direction of flow. So as the flow rate increases the thickening effect will be reduced, up until some flow rate where all chains are aligned. An increase in the flow rate above this value will not cause any further reduction in thickening amount by this alignment mechanism.

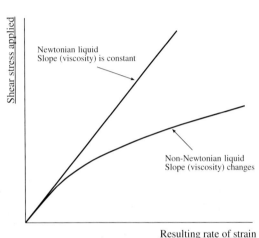

Figure 4.2

This phenomenon is called temporary viscosity loss or (somewhat misleadingly) temporary shear of the viscosity index improver. It's temporary because, once the oil returns to rest, the chains will randomise their orientation again and the original degree of thickening will be restored.

As explained in the previous chapter, the definition of viscosity is the ratio of applied shear to the resulting rate of strain. The viscosity of liquids such as water or

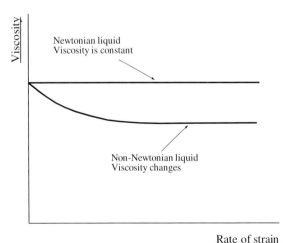

Figure 4.3

The roles of lubricant components

mineral oils does not change with rate of shear. A plot such as Figure 4.2 will be a straight line. However, if we make the same plot for an oil that has been thickened with VII we will not get a straight line. As has been explained, the viscosity varies with movement of the fluid or rate of strain and, hence, the slope of the plot will also depend on the rate of strain. The outcome is a curve which straightens out at higher shear rates, once the polymer molecules are all aligned with the flow (but see next paragraph). Liquids which give a straight line – that is, liquids whose viscosity does not vary with shear or strain – are called Newtonian liquids. Perhaps obviously, those that do vary are called non-Newtonian (Figure 4.3).

As well as the temporary effect that I have just explained, exposure to shear can also permanently reduce the viscosity of a liquid that has been thickened with polymer chains. Depending on the chemical makeup of the polymer, shear can physically break chains, making them shorter but more numerous. The net effect of such a change is a reduction of thickening effect, a single long chain causing more interference to the flow of the base oil than two shorter ones. Different chemical types of polymers that are used as VIIs have different levels of shear stability or resistance to mechanical fracture by shear.

If a multigrade oil is formulated with a non-shear stable VII, its viscosity will reduce rapidly (and permanently) when put into service. What started as an SAE 40 grade (see Table 3.1) may shear down to an SAE 30 or less. It's an unfortunate reality that, in general, polymers that give the greatest thickening effect are also the least shear stable. Accordingly, there's a cost incentive for the formulator of a lubricant to use a high thickening power/low shear stability polymer to produce a given multigrade oil, rather than to use a more shear stable one. The formulator will be able to use less polymer and, polymer being more expensive than base oil, the cost of the product will be less than one made using a polymer with a higher shear stability.

Shear stability index

In the text I've explained that multigrade oils can vary in their degree of shear stability – different polymeric VI improvers having different resistances to mechanical shearing action. The shear stability level is often expressed by a measure called the shear stability index.

When a VI improver is added to a base oil it increases the oil's viscosity. If its original viscosity was Vb then, after addition of the VI improver, it will increase to V. After a period of shearing, the viscosity will drop to a value Vs, with Vs lying somewhere between V and Vb. The shear stability index (SSI per cent) is defined as:

$$\text{SSI per cent} = \frac{V - V_s}{V - V_b} \times 100$$

Expressed in words, the shear stability index is the fraction of the viscosity contributed by the viscosity improver additive that is lost during shearing, expressed as a per cent.

You will note that if an oil is perfectly shear stable, then its SSI will be 0 per cent (there will be no difference between V and Vs). There will be no loss due to shear. If it is totally shear unstable, its SSI will be 100 per cent (the thickening effect of the polymer will be completely destroyed and Vs will be equal to Vb).

Which Oil?

> ### The Mini Minor and multigrade oils
> In the UK (and subsequently in Europe) the arrival of the Mini Minor in 1959 was ultimately a boon for multigrade oils and relatively shear stable viscosity index improvers.
>
> The gearbox in the Mini was in the sump of the engine, and relied on the engine oil for its lubrication. To control gear wear, the oil had to be relatively viscous. However, a viscous oil gave difficulty with gear changes, particularly when the sump was cold. It also was not ideal for the lubrication needs of the engine. The solution was to use a 20W/50 multigrade. The SAE 50 viscosity when hot was excellent for the gears and satisfactory for the engine while the SAE 20W viscosity at low temperature gave acceptable shift characteristics. However, to be successful, the oil had to be relatively shear stable as it was operating in a very high shear environment.

The United States has suffered from this situation probably more than any other region of the developed world. In the 1960s and 1970s, low shear stability multigrade oils were virtually the norm in the US market. While satisfactory for the vast majority of American vehicles (which were generally understressed V8s), they did cause some problems in higher performance engines. This has led to many misconceptions about the performance and the inherent attributes of multigrade lubricating oils.

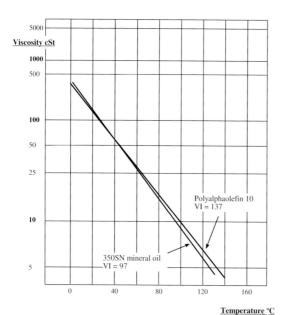

Figure 4.4

By contrast, Europe (including the UK) was largely spared such low shear stability products. This was probably in part due to the greater technical emphasis that tends to prevail in Europe on the whole, the generally smaller and more highly stressed engines used there, and also the arrival, early in the commercial deployment of these products, of the Mini Minor (see panel).

Also, until the late 1980s, the standard US auto industry specifications for engine oil viscosity did not include any stringent tests or requirements for shear stability. Hence, the user

The roles of lubricant components

was entirely in the hands of the oil marketing companies. It was simply not possible to judge from the labelling of a lubricant whether or not it would stay in its original viscosity grade during use. Again, Europe was much better in this regard, and a requirement for shear stability was introduced in European specifications some ten years earlier. The situation has been helped in recent years by the move from Group I to Group II and particularly Group III base oils (see panel in Section 4.1). These have higher inherent viscosity indices, so do not need as much polymer added.

4.1.2 Synthetic base oils

As was mentioned in Chapter 3, some synthetic base oils have inherently high VIs. Their viscosity changes less with temperature than oils derived from natural mineral oil. The use of such oils can avoid or reduce the need to add another substance to improve the VI of a lubricant.

Commonly used synthetic base oils that have high VIs are ones based on polyalphaolefins (PAOs). A typical PAO has a Viscosity index (VI) of greater than 130, while a typical Group I paraffinic mineral oil has a VI in the range 95 to 100 (see the panel on Base Oil Groups at the start of this chapter). Figure 4.4 shows a comparison.

However, not all synthetic base oils have high VIs. Another commonly used group are the synthetic esters and, as a generalisation, their VIs are no higher than those of mineral oils. They are used because of other valuable characteristics.

4.2 Reduction of oxidation susceptibility

In the previous chapter I explained that hydrocarbon-based oils are vulnerable to chemical reaction with oxygen. The rate of reaction is negligible at ambient temperatures, but becomes noticeable above about 120°C, and rapid above 170°C.

Oxidation is a complex chemical reaction. It is started by what is called an initiator (a type of chemical entity called a free radical) and proceeds via a chain reaction. If an oil is exposed to oxygen at a high temperature there is an initial period where, apparently, nothing happens. Then oxidation begins and proceeds rapidly. The delay is while the initiators are forming. Once they reach a certain critical level, the chain reactions begin and a 'runaway' situation develops.

Given this mechanism, it's not surprising that two types of anti-oxidants have been developed. One type stops the build up of the initiators. The other stops the chain

---

Oil drain

All instructions on engine oil change advise you to warm the old oil thoroughly before draining. This is very good advice. Given the way in which oxidation proceeds chemically – with the need for precursors or initiators to be generated – it is essential to remove as much as possible of the old oil before adding the new. Any oil left behind will already contain the precursors and so its presence will allow oxidation of the new oil to commence more rapidly than in the case of a complete drain. This will lead to a faster depletion of the antioxidant additives and, ultimately, a shorter oil life. The hotter the oil, the lower its viscosity and the quicker and more complete the drain. For the same reason, the oil filter should preferably be changed at every oil drain or, at minimum, drained of all old oil.

Which Oil?

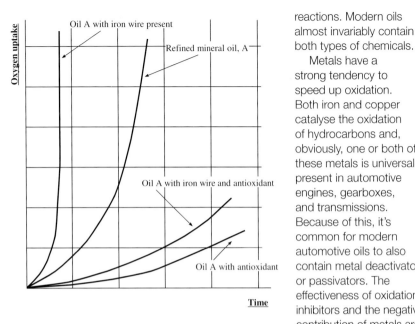

Figure 4.5

reactions. Modern oils almost invariably contain both types of chemicals.

Metals have a strong tendency to speed up oxidation. Both iron and copper catalyse the oxidation of hydrocarbons and, obviously, one or both of these metals is universally present in automotive engines, gearboxes, and transmissions. Because of this, it's common for modern automotive oils to also contain metal deactivators or passivators. The effectiveness of oxidation inhibitors and the negative contribution of metals are shown in Figure 4.5.

4.3 Dealing with mixed or partial lubrication regimes

The approach that is used to handle the mixed lubrication regime (see Section 3.5) is to include in the lubricant, by means of chemical additives, molecules of a type that will tend to be attracted to the surfaces involved and form a protective film on them. These molecules may be either oil molecules or non-oil molecules. If chosen correctly, non-oil molecules can form stronger protective films than the oil ones.

The types of oil molecules that are attracted to metal surfaces are ones with some polarity. Polarity means that, while each molecule is electrically neutral, the electrons in it are not uniformly distributed. One portion of the molecule is positively charged and another portion is negatively charged. Such molecules are attracted to metals because metals also have a non-uniform distribution of electrical charges at an elemental level.

Polar oil molecules may form a layer on a metal surface which is several molecules thick. This layer remains in place even if the bulk oil drains away or is squeezed out, and hence protects the surface. However, the force of attraction is relatively weak and so such films give protection only under moderate loads (the right-hand end of the mixed or partial lubrication region on the Stribeck curve; see Figure 4.6). The presence of a film on the metal surfaces also reduces the frictional force in this zone, and this is important for fuel efficiency (see Section 4.6, following).

The roles of lubricant components

These types of additives are called friction modifiers or reducers. They also give modest wear protection.

Moving away from oil-type molecules, it's possible to find or synthesise non-oil substances that will form a soft chemical layer on metal surfaces. Such chemically-bonded layers can resist higher loads than the oil ones that attach by polarity alone. These types of substances are called anti-wear additives and they can give protection through the entire mixed lubrication region (Figure 4.6).

The challenge is to find chemicals that can form films and that are also completely stable in the lubricant – under all of the conditions that it will encounter – while not bringing any negative effects. In reality, only a quite limited number of substances has been found to be appropriate, and they usually contain at least one of the chemical elements phosphorus or sulphur. The most widely used has been zinc dithiophosphate (ZDTP).

4.4 Dealing with boundary lubrication

In the boundary regime (Figure 4.6), no oil film is present, and surface films – either polar oil or chemical bonding-based – cannot be maintained. Surface-to-surface contact cannot be avoided. If such a situation exists in a mechanical device, and it were allowed to persist during operation, then disastrous rates of wear would result. Also, the wear products – small abraded pieces of the surfaces involved – would contaminate the lubricant and exacerbate the wear, not only on the surfaces where the boundary problem exists but on all other lubricated surfaces.

So, what can be done if, as in a hypoid differential, operation in the boundary region cannot be avoided? Protective films that are deposited from polar oily molecules or other non-oil molecules will not be strong enough to resist the loads and will be worn through, particularly at the critical high points of the contacting surfaces.

If the heavily loaded surfaces are made of iron or steel – which is almost universally the case in motor vehicles – the approach adopted is to add chemicals

---

Polar base oils

Many of the molecules that make up conventional mineral base oils have some inherent polarity. They tend to adhere to or wet metal surfaces. A typical mineral base oil contains many hundreds of different types of molecules, the types differing to a degree with the original source of the crude oil from which the base oil was refined, the refining process adopted, and the particular base oil cut (or viscosity grade).

Synthetic base oils consist of a much more limited range of molecules. Some types have inherent polarity and others have very little. Formulators have to be aware of this and allow for it when developing oils based on synthetic base stocks.

At least one very well known marketer has taken this use of polar molecules to a logical conclusion. It has an engine oil that it promotes as clinging to metal surfaces and not fully draining away when the engine is shut down, thus providing a degree of protection on start-up.

Which Oil?

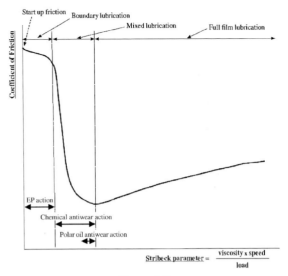

Figure 4.6

to the lubricant which, at very high temperatures, react with the metal surfaces themselves and soften them. The result is that, instead of binding and welding together, the surfaces flow or shear, and hence slide over each other. Rather than depositing a protective film on the surface, the surface reacts with the additive to form its own film. Some sacrificial wear occurs all the time that the device is in operation but, because the extent to which the softening reaction occurs is literally only a few atoms deep and, as explained below, is also very limited in surface area extent, the rate of wear is very low.

While the approach has some similarity to that for the anti-wear additives covered in Section 4.3, there are crucial differences. As explained, anti-wear additives in general operate by depositing a soft chemical layer on the surfaces. Boundary regime additives – which are known as extreme pressure or EP additives – react with the metal surface itself and, crucially, do so only at very high temperatures. These high temperatures occur only at the very localised, contacting high spots of the two surfaces, and these high spots are microscopic in extent, at least initially. Thus, the only portions of the surfaces softened are the contact portions, and it is only these microscopic high spots that are sacrificed. There is not a continuous covering of softened metal over the surfaces, nor a mass loss of metal.

Because there is no protective film, it also follows that the film cannot be breached at high loads, as can happen with anti-wear films. It's the high load that generates the high localised temperature, which in turn triggers the localised chemical reaction, which in turn softens the metal involved in the contact point. The high contact point flows and is smeared out, and wear is avoided. Thus, the EP approach can handle higher loads than the film approach, with the reaction and softening happening exactly where it's needed, and only where it's needed. It gives protection in the boundary region of the Stribeck curve (Figure 4.6).

Another helpful outcome is that, after a period of initial operation, all of the microscopic high spots will have tended to have been softened and smeared out. The surfaces will have become smoother, meaning that they can be kept separated by a thinner oil film than was the case at initial start-up. The load regions over which

The roles of lubricant components

full-film and/or partial film lubrication can be maintained will have been extended. This phenomenon is known as running in.

EP additives are almost universally based on sulphur. In fact, the original approach was simply to mix powdered, elemental sulphur into the oil. However, the sulphur tended to settle out and fall to the bottom, so there was a rapid development of sulphur-containing chemicals that were soluble in mineral oils but whose make up was such that the sulphur was readily available for other reactions. Such sulphur is called 'active.' Many of these compounds were originally based on sperm whale oil as it had good solubility in mineral oils and could react readily with sulphur and act as its carrier.

Sulphur reacts with iron only at very high temperatures. However, it reacts with softer metals, such as copper, at much lower temperatures. If copper comes into contact with a sulphur-based EP additive at temperatures over about 100°C it will be corroded. Copper is present in many of the components used in motor vehicle engines and gearboxes (bearings, bushes and some synchronisers) and so the use of EP additives based on active sulphur is limited. In motor vehicles it is usually confined to differentials.

4.5 Dealing with emissions

Since the 1970s, in parallel with the drive to better fuel economy, there has been the need – usually legislated – to reduce the pollutants emitted by motor vehicles. This has led to changes in the fuel (such as the elimination of lead from gasoline and the reduction of sulphur in both gasoline and diesel fuel), changes in vehicle hardware (such as the installation of catalytic converters and/or particulate filters in exhaust pipes), and changes to combustion processes (such as the use of exhaust gas recirculation). Passive or static emissions have also had to be reduced, so carburettor, fuel tank, and crankcase vents all have had to be fed to the engine's air inlet, rather than discharged to atmosphere.

Many of these changes have had an impact on the engine lubricant. Perhaps the largest one was the introduction of catalytic converters. As was explained in Section 4.3, modern engines require the presence of an effective anti-wear additive in the lubricating oil. The most widely used chemical substance employed to give this attribute is zinc dithiophosphate (ZDTP for short). It also acts as an effective antioxidant.

Catalytic converters facilitate the conversion of pollutants, such as carbon monoxide, to less noxious substances (in this case, to carbon dioxide). They contain metals such as platinum, deposited on a highly porous ceramic or metallic substrate to give a very large surface area. The platinum catalyses (speeds up) the chemical conversion of the pollutants in the engine's exhaust gas as they pass through the converter.

As well as the by-products that arise from the combustion of the fuel in the engine, the exhaust gases also contain a small amount of by-products from the combustion of the lubricant. This situation cannot be avoided. To lubricate the cylinder bore/piston ring interface, a film of oil must be deposited on the bore, and

Which Oil?

some of this will be burnt in each combustion cycle. One of the products of this combustion will be compounds such as phosphates (formed by the combination of phosphorus and oxygen). The phosphorus comes from the ZDTP anti-wear/ antioxidant additive, and the oxygen comes from the combustion air.

These phosphates (and some other phosphorus-based compounds) have a strong tendency to deposit on the platinum-coated surfaces in converters, rendering them less functional as a catalyst. This is known as poisoning of the catalyst and, if it occurs, exhaust emissions will rise. Because of this effect, the amount of phosphorus that can be present in an engine oil intended for the lubrication of a vehicle with an exhaust catalyst has been restricted by mutual agreement between the car makers and the major formulators of lubricants, albeit with the car makers applying the pressure. In the US, a binding restriction to 0.12 per cent maximum phosphorus content was first introduced in 1993 (for some viscosity grades, the ones recommended by the car manufacturers for the then-new cars), although many oil suppliers had moved their formulations earlier than this date. There was a further reduction to 0.10 per cent in 1996 and then to 0.08 per cent in late 2004.

In Europe, there was no mandatory move in the late 1990s to reduce phosphorus. With the smaller, more efficient engines, catalytic convertors were not needed to the extent that they were in the US. Then, rather than introducing chemical limits across the board, the Europeans introduced a separate family of engine oils in 2004 designed to be compatible with exhaust emission treatment devices in exhaust pipes or tailpipes. These are the ACEA 'C Sequence' oils, and they are explained in the next chapter (Section 5.1.2).

It should be noted that, when a limit was introduced on the amount of phosphorus that could be present in the engine lubricant, there was no reduction in the pass levels in the various tests, including engine wear tests, that an oil must achieve to be able to claim a particular performance specification. Nor was there any reduction in the severity of the tests. Rather, they became more demanding.

Reduction of phosphorus was really a reduction in zinc dithiophosphate. This additive gives the oil both an enhanced ability to resist oxidation and also anti-wear

---

Chemistry of ZDTP and its anti-wear action

Zinc dithiophosphate (ZDTP) is a complex molecule, and variations in its exact chemical make up are possible. These variations, which are controlled deliberately by the manufacturer, affect its performance as an antioxidant and its performance as an anti-wear additive.

Depending on the starting raw materials, the ZDTP made can be either of a primary form or a secondary form. An explanation of these terms is beyond the scope of this book but, if you have an interest and some understanding of chemistry, look up the definition of primary and secondary alcohols in a chemistry textbook or on the internet.

The primary forms of ZDTP have better high temperature stability but a lower anti-wear performance than the secondary forms. Oil blenders use a mix but the ratio of primary to secondary tends to be higher in engine oils formulated for heavy-duty diesel engines (as used in large trucks), whereas in oils for gasoline engines the opposite is true.

The roles of lubricant components

performance. Antioxidants that do not contain phosphorus are quite readily available. Suitable phosphorus-free anti-wear chemicals are far more limited.

The first response of the oil industry (in 1993 and 1996) was to use what are known as 'more active' versions of zinc dithiophosphate. Technically, they are based on secondary rather than primary zinc chemistry (see the panel). They give a higher level of anti-wear performance for the same amount of additive. Hence, to maintain the current performance level, less additive could be used, resulting in less phosphorus in the oil. Any reduction in resistance to oxidation was compensated for by using non-phosphorus antioxidants.

The introduction of the 0.08 per cent limit on phosphorus in 2004 could not be met by such a move. New, non-phosphorus anti-wear additives had to be used, and more non-ZDTP antioxidants. However, the actual anti-wear and antioxidant performance of the oils, as measured by standard engine tests, was not allowed to be compromised.

The dates and values listed above for phosphorus restrictions are those that prevailed in the US. In general, the rest of the developed world followed (and sometimes led). However, the latest reduction to 0.08 per cent maximum has not been adopted universally. More details are given in the next chapter.

4.6 Friction modification

The phrase 'friction modification' means the use of additives to deliberately change the coefficient of friction of the lubricating oil. It needs to be understood that a particular value of coefficient of friction exists only in relation to the interaction of the oil with a particular solid. It is not a property of the oil that can be measured or quoted in isolation.

Historically, additives have been used to modify the frictional characteristics of the oils used in automatic transmissions. This has been done to control the gearshift characteristics, to make them smoother. That application is dealt with in Chapter 6. In more recent times – since the mid 1990s – additives have been used in some gasoline engine oils to reduce the frictional losses in the engine, to improve fuel consumption.

5. Engine lubrication

The functions that, ideally, an engine lubricant should perform are summarised in Figure 5.1. The first lubricants for motor vehicle engines simply looked after wear and friction. They functioned successfully only between those surfaces where a full film of lubricant could be maintained. Control of the condition of the lubricant itself, and the elimination of contaminants from combustion and wear, were handled by replacing the lubricant at frequent intervals.

To a large degree, the motorist had no choice in this. Early engines consumed oil at a prodigious rate – very early ones were lubricated, at least in part, by direct oil feed into the cylinders, a total loss system – and so top up was almost continuous. Even with this rate of consumption, recommended oil drain intervals in closed lubrication systems were very short by modern standards – in the order of a few hundred miles or kilometres of travel, or a few days of operation. This was necessary to remove solid contaminants.

The evolution of lubricants since that time has been mainly on three fronts – increasing the time interval between oil changes, expanding the temperature range over which the lubricant could maintain a full-film between moving surfaces, and introducing chemicals to reduce wear in those parts of the engine where a full-film could not be maintained. Some of these changes were driven by the evolution of the engines (higher compression ratios and power outputs, giving greater loads on moving surfaces), and some by consumer preference (eg less oil consumption and longer drain intervals). Table 5.1 summarises this evolution, from an additive perspective, by approximate date.

If you own an older vehicle, in order to choose the correct lubricant for your engine from the modern products that are on offer, you need to be able to answer two basic questions. First, what lubricant performance level (if any) did the engine manufacturer specify when it was new? Second, what viscosity did it specify?

5.1 Performance level

Fortunately, for many years now, engine manufacturers in most of the developed world have used a common language to

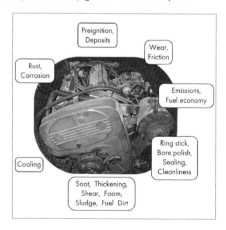

Figure 5.1 Roles fulfilled by engine lubricant.

Engine lubrication

specify the required performance level of engine lubricants, albeit a language that has differed in detail and vocabulary between Europe and the US.

5.1.1 United States

Beginning in 1947, the combined US oil companies – under the auspices of the American Petroleum Institute (API) – began to specify the performance level of appropriate lubricants for gasoline (petrol)-powered four-stroke engines. Originally, three performance levels of engine oils were defined: Regular Type, Premium Type and Heavy-duty Type. The performance level met by a particular oil depended on the presence or absence of chemical additives. In general, the Regular oils were straight mineral oils, the Premium oils contained some oxidation inhibitors, and the Heavy-Duty oils contained antioxidants plus, in some cases, limited amounts of detergent-dispersant additives. No differentiation was made between oils for gasoline engines and those for diesels.

The approach was extended and improved in 1952, and further revised in 1955. It was again based on three performance or duty levels, but the different needs of gasoline and diesel engines were recognised. The three performance levels for oils for gasoline engines were designated ML (Motor Light), MM (Motor Medium), and MS (Motor Severe). The three for diesels were DG (Diesel General), DM (Diesel Medium), and DS (Diesel Severe). Sometimes, the MS (Motor Severe) oils were still referred to as HD (Heavy-duty) oils.

The system was modified again in 1960. It became based not on chemistry but on the performance of an oil in a series of actual engine tests, the specific tests being selected and run by the API. However, many engine manufacturers still specified additional tests, some of them in-house and proprietary. This restricted the availability of oils for consumers, limiting them to certain brands in some circumstances.

In response to this situation, in 1969/70 the combined US oil and automotive industries – assisted by the SAE – developed a co-operative system of performance classification. To meet a particular performance specification a lubricant had to pass a series of severe engine tests which were developed and selected by the motor manufacturers to cover the full range of their performance needs. The specifications are known as Service Classifications and, for gasoline engines, they are identified by

Year	Anti-foam	Pour point depressants	Anti-oxidants	Zinc/phosphorus antiwear	Anti-corrosion	Detergents	Dispersants	Viscosity index improvers
1925	✓	✓						
1930	✓	✓	✓					
1935	✓	✓	✓	✓	✓			
1960	✓	✓	✓	✓	✓	✓		✓
1965	✓	✓	✓	✓	✓	✓	✓	✓

Table 5.1 Approximate year of commercial introduction of additive to engine oil.

Which Oil?

the letter 'S,' followed by a second letter (eg SA, SB, etc). The objective was – and still is – to ensure that oils are available to meet the operational needs of current engines, and to give the owner of the engine a method of knowing that a given oil product is suitable for it, without being tied to a particular product or brand. It's essentially the same system that exists today.

When it was introduced in 1971, the classification system was applied retrospectively to the oils that had been used and marketed in earlier times. Classification SA was defined as additive-free mineral oil, typical of the engine oils that had existed in 1930 and earlier. These were the equivalent of what had been called ML or Regular oils. Classification SB applied to engine oils that had some degree of antioxidant, anti-wear and anticorrosion protection, typical of the oils available in the period from 1931 to 1963. They did not contain detergents. They were similar to what had been called MM or Premium oils.

SC was the classification that applied to the type of oils that had been introduced by the oil industry in 1964 under the label MS. That had been the first time that the specification of engine oil performance had been based on performance in engine tests rather than on chemistry. Also, detergents had been necessary to pass some of the tests involved.

Finally, to complete the retrospective classifications, SD applied to the oils that had been introduced in 1968 to meet revised, more severe engine tests. These oils had also been called MS. It was the fact that, although the testing severity was increasing over time the performance description was staying unchanged, that prompted the move to the new classification or 'numbering' system. Under the new system, it would be possible to differentiate an old specification oil from a new one.

SA to SD were all retrospective performance classifications. The actual new classification that was introduced with the modified system in 1971 was SE. It ran from 1972 to 1979.

As I have mentioned, because engine performance and driving demands have generally increased over time, the engine oil performance specification also had to increase to keep up. Rather than simply changing the specification tests or limits that a lubricant must meet in order to qualify, the approach (since 1971) has been to leave the current classification unchanged and introduce a new, more demanding one. Thus, at any point in time, there is a current classification and a continuous series of historical classifications. From the introduction of the new classification system with SE in 1972, we are now – in 2011 – up to classification SN. However, the SAE does ultimately declare older classifications to be obsolete. Currently, SG and all earlier classifications are technically obsolete. Amongst other difficulties, the engine test equipment used is no longer available.

When a new classification is introduced, it's a requirement that the oils that meet it are also 'backwards compatible.' That is, they must also be suitable for use under the classification that they are superseding. This is to ensure that the new oils will be suitable for use, not just in the new engines or service demands for which they've been developed, but also in the existing engines in the marketplace. Table 5.2 lists the API Service Classifications, in historical (date) order.

Engine lubrication

Table 5.2 is a very cryptic summary and does not do justice to the strenuous testing regimes that were and are necessary to meet the various classifications. However, it will suffice for our needs.

It should be noted that the service classifications generally do not specify required viscosity for engine oils. Up until the introduction of SG in 1987, viscosity was not really mentioned or specified in them at all. In SG a shear stability test was introduced – the L38 stay-in-grade requirement. However, it was not (and still is not) very severe. The underlying concept is that the appropriate viscosity to use in an engine is determined to a great degree by the ambient temperatures in which it will be operating. The particular engine supplier must make the recommendation on this. However, since the early 1990s, the SAE engine oil viscosity classification system has included a high temperature/high shear specification requirement (see Section 5.2).

Although the fundamentals of the US engine oil performance classification system have remained unchanged since 1971, during the 1980s – as the motor industry came under great pressure on emissions and fuel economy – it felt that its partner, the oil industry, was being too slow in evolving the engine oil performance specification. The joint approach came under real threat. The American Automobile Manufacturers Association (AAMA) got together with the Japan Automobile Manufacturers Association (JAMA) to form the International Lubricant Standardisation and Approval

API classification	Applicable period	Properties specified
SA	Up to 1930	Mineral oils with no additives.
SB	1931-1963	Oxidation resistance. Corrosion prevention. Wear control.
SC	1964-1967	Based on engine tests. As for SB but with detergents/dispersants. Control of coking deposits. Control of sludge.
SD	1968-1971	As per SC but more severe tests, hence higher levels of additives.
SE	1972-1979	As per SD but more severe tests and higher levels of additives. Performance in stop/start operation.
SF	1980-1988	As per SE but more severe tests and higher levels of detergency. Some suitability for diesel engines.
SG	1989-1993	As per SF but more severe and some additional tests. Higher dispersancy. Limit on phosphorus content (0.12%) and introduction of modest shear stability test. Limits on evaporation loss, filterability, foaming and flash point (flammability).
SH	1993-1996	Similar to SG but with requirement that all testing be done in accord with new CMA code.
SJ	1996-2001	As per SH but lower evaporation loss and reduced phosphorus (0.10%). New high temperature deposit and low temperature gelation tests.
SL	2001-2004	As per SJ but more severe tests. May also be energy (fuel) conserving.
SM	2004-2010	As per SL but more severe tests. Reduced phosphorus (0.08%).
SN	2010-	As per SM but more severe tests. Still 0.08% max phosphorus for low viscosity grades, but relaxed for higher grades.

Table 5.2 History of API engine oil classifications.

Which Oil?

Association (ILSAC). It began to develop oil specifications for gasoline engines under its own banner, publishing the first one – GF-1 – in 1993. The API specification SH, which was also issued in 1993, mirrors it. We are now (in 2011) up to GF-5.

The ILSAC specifications apply only to the new passenger car engines current in the US at the time the specification is developed. They reflect the main concerns of the automotive industry at that time, and may not be totally appropriate for older cars. For example, they are very concerned with fuel economy and emissions. The oils specified tend to be the lower viscosity ones and a limit is set on phosphorus content. They, perhaps increasingly, will generate a category of oils somewhat akin to the 'C' oils in Europe; the so-called catalyst-compatible oils (see Section 5.1.2, following). However, the ILSAC specifications certainly do heavily influence the API oil performance classifications.

This influence is shown by the most recent API engine oil performance specifications, SM and SN. To reduce exhaust catalyst poisoning, the maximum amount of phosphorus allowed in an oil that meets these specifications is 0.08 per cent, down from 0.10 per cent in the previous specification, SL, (although SN does not impose this restriction on higher viscosity oils).

5.1.2 Europe

If the US engine oil classification system appears complex, it is simplicity itself when compared with the current situation in Europe. This was not always the case. Up until 30 years or so ago the European approach was relatively simple. Car manufacturers either recommended particular oil brands or used terminology based on the US system.

Control of engine oil performance claims

While the tests and physical and chemical properties that an engine oil must meet to be able to claim a particular performance level (SG, SL, etc) are fully defined, historically, the oil and car industries had little or no control over what individual oil marketers claimed for their products. All testing was private.

Since 1993, the API has run a performance claim licensing system. Marketers submit test data and a fee to the API and, in return, if the test results are accepted, the marketer can use the API's 'donut' (the logo on the left, below) on the packaging. The presence of the donut shows that there has been an independent verification of the performance level claimed for the particular oil.

To ensure that oils in the marketplace that display the donut continue to meet the test results submitted originally, the AIP also has a systematic programme of random sampling and testing of retail products.

The AIP also administers the parallel 'starburst' certification mark for oils that also meet the ILSAC requirements.

Engine lubrication

This began to change in the 1970s in response to pressures on emissions and the politics of the European Union. However, it certainly was a fact that European engines were very different in average characteristics when compared with US ones. They were much smaller, faster revving, and had generally higher power outputs for a given engine capacity. On average, they were more highly stressed.

The first step in recognition of this was to add some specifically European engine tests onto the US tests. The changes began with individual car manufacturers, but the situation was soon taken under the wing of a body called the Co-ordinating European Council for the Development of Performance Tests for Lubricants and Fuels (the CEC), in conjunction with the Committee of Common Market Automobile Constructors (the CCMC). Beginning in 1976, the specifications were issued as CCMC specifications, and the version that was current in the first half of the 1990s is summarised in Table 5.3. The outcome was that, in some respects, the European specifications were more severe than their US counterparts.

Political difficulties caught up with the CCMC and, in 1996, it was replaced by the Association des Constructeurs Europeens d'Automobiles (the ACEA). Also, the European automobile industry was going through major structural change, with three German manufacturers – Daimler-Benz (Mercedes), Volkswagen-Audi, and BMW emerging as the dominant players and the technology leaders.

This competitive environment was not very conducive to the development of common engine oil performance specifications, particularly with the oil industry being somewhat sidelined. There was a natural reluctance for any one of the dominant car manufacturers to endorse an oil specification that was based, in any significant part, on a standard engine test provided by one of its competitors. Also, in fairness to the manufacturers, they tended to see the lubricant as an integral part of the design of the engine. They wanted to have more control over it, the same sentiment that led to the formation of ILSAC in the USA (see Section 5.1.1).

So, as the politics (both between manufacturers and between member states)

CCMC specification	Period	Description
G1	-1989	Corresponds approximately to API SE but with three additional European engine tests
G2	-1990	Corresponds approximately to API SF but with three additional European engine tests
G3	-1990	Also corresponds approximately to API SF but with three additional European engine tests and higher requirements re oxidation and volatility. Applies to low viscosity oils.
G4	1990-1995	Corresponds approximately to API SG but with additional European black sludge and wear tests
G5	1990-1995	Corresponds approximately to API SG but with additional European black sludge and wear tests and greater demands than G4. Applies to low viscosity oils.

Table 5.3 European engine oil specifications in the 1990s.

Which Oil?

intensified, the 'centralised' specifications became more and more complex as they tried to accommodate all parties. However, even with this complexity, the three major manufacturers insisted that oils for the factory-fill of their engines must also meet specific in-house tests and approval processes. Unfortunately for us, they also abandoned the sequential numbering system. Instead, given categories now evolve in performance level over time, making it more difficult to interpret older information in today's context. The most recent update was in 2010.

Given this background, it's not surprising that it's difficult to give a succinct

Sequence	Description/application
A/B	Gasoline (petrol) and light diesel engined cars and vans
A1/B1	Oil for use in engines specifically designed to be capable of using low friction low viscosity oils with a high temperature / high shear rate viscosity of 2.6 to 3.5 mPa.s. These oils may be unsuitable for use in some engines.
A2/B2	No longer defined
A3/B3	Stable, stay-in-grade oil for use in high performance engines and/or for extended drain intervals where specified by the engine manufacturer, and/or for year-round use of low viscosity oils, and/or for severe operating conditions as defined by the engine manufacturer.
A3/B4	Stable, stay-in-grade oil intended for use in high performance gasoline and direct injection diesel engines, but also suitable for applications described under A3/B3.
A5/B5	Stable, stay-in-grade oil intended for use at extended drain intervals in high performance engines designed to be capable of using low friction low viscosity oils with a high temperature / high shear rate viscosity of 2.9 to 3.5 mPa.s. These oils may be unsuitable for use in some engines.
C	Catalyst compatibility oils for gasoline (petrol) and light diesel engined cars and vans
C1	Stable, stay-in-grade oil intended for use as catalyst compatible oil in vehicles with diesel particulate filter (DPF) and/or two way catalyst (TWC) in high performance engines requiring low friction, low viscosity, low sulfated ash/phosphorus/sulfur (SAPS) oils with a HTHS higher than 2.9 mPa.s. These oils will increase the DPF and TWC life and provide fuel economy benefit. Warning: these oils have the lowest SAPS limits and may be unsuitable for use in some engines.
C2	Stable, stay-in-grade oil intended for use as catalyst compatible oil in vehicles with DPF and TWC in high performance engines designed to be capable of using low friction, low viscosity oils with a HTHS higher than 2.9 mPa.s. These oils will increase the DPF and TWC life and provide fuel economy benefit. Warning: these oils may be unsuitable for use in some engines.
C3	Stable, stay-in-grade oil intended for use as catalyst compatible oil in vehicles with DPF and TWC in high performance car and light van diesel and gasoline engines. These oils will increase the DPF and TWC life. Warning: these oils may be unsuitable for use in some engines.
C4	Stable, stay-in-grade oil intended for use as catalyst compatible oil in vehicles with DPF and TWC in high performance car and light van diesel and gasoline engines requiring low SAPS oil with HTHS higher than 3.5mPa.s. These oils will increase the DPF and TWC life. Warning: these oils may be unsuitable for use in some engines.

Table 5.4 Current European engine oil performance classifications.

Engine lubrication

summary of the current engine oil specification system in Europe. Broadly, it has four streams or sequences, designated A, B, C and E. They cover oils for use in service (as opposed to factory-fill). The A sequence is for gasoline (petrol) engines, the B sequence is for light duty diesel engines (as found in a large number of cars in Europe), C is for catalyst and exhaust gas filter compatible oils, and E is for heavy-duty diesel engines, as found in trucks, buses, etc.

Since 2004, the A and B streams have been combined. That is, oils in this combined stream must be able to satisfy both gasoline and light diesel engines. This is in recognition that both types of engines are widely used in the passenger car fleet in Europe and, hopefully, to reduce both the necessary range of oil products in the marketplace, and the scope for consumer confusion. Within each stream or sequence there is also a performance or application hierarchy.

Also, as already noted, the tests that must be met in order to meet a particular performance specification evolve over time. An oil of A3 specification in 2002 may not necessarily pass the A3 specification tests today. Probably of more relevance to us is the fact that an A3 oil of today is not necessarily the same performance level as one of 2002. It will be of higher performance, and other characteristics may have changed.

Finally, in Europe since 2004, an additional performance sequence has existed to define oils that are specifically formulated to be compatible with a variety of exhaust gas emissions treatment systems, including catalysts. They are the 'C' sequence oils (C1, C2, etc). They have low phosphorus levels and are also low in some other chemicals, such as sulphur.

Table 5.4 reproduces a summary of the current European engine oil specifications that is published by ACEA. I've not included the heavy-duty diesel sequence, E. You will note that there are some apparent gaps as some levels of the hierarchies have been dropped in recent years. You will also note that the majority of the definitions or

---

US versus European engine oils

Since the 1980s – when they were based on the US API 'S' system of performance classifications, plus a few additional tests – European requirements for engine oils have been higher than those in the US. However, what was a modest performance difference back then has become a much wider gap since the late 1990s/early 2000s. Today, even allowing for the influence of ILSAC and its GF specifications in the US, the gap is significant.

It's difficult to make a simple yet fair and meaningful comparison. I have attempted one in Table 5.5. I have listed five engine/oil life performance areas and two environmental ones, and awarded points for relative performance in each area. A score of 5 is outstanding and 1 is poor. An absence of a score against a particular feature means that the oil concerned does not claim to address this item. The higher the total score, the better the oil for a modern car engine.

The European oils do not receive a score for catalyst compatibility. Oils from some manufacturers do have phosphorus limits but, from a centralised specification perspective, the 'C' sequence oils are specifically designed for compatibility with exhaust gas treatment systems, including catalysts.

Which Oil?

Performance area	US (API SN) Oil	Europe (A1/B1) oil	Europe (A3/B3) oil	Europe (A3/B4) oil	Europe (A5/B5) oil
Piston deposits	4.0	3.5	3.5	4.5	4.5
Wear	2.0	4.0	4.0	4.0	4.0
Sludge	1.5	4.5	4.5	4.5	4.5
Soot thickening		3.0	3.0	3.0	3.0
Oxidation	3.5	2.0	2.5	2.5	2.5
Catalyst compatibility	4.0	-	-	-	-
Fuel economy	3.5	3.5	-	-	3.5
Total	18.5	20.5	17.5	18.5	22.0

Table 5.5 Performance comparison of US and European engine oils.

descriptions are not very helpful. They tend to be of the form: "This oil is suitable for use in those engines for which it is suitable." Table 5.5 gives a simplified comparison of the various classifications and compares them with the US. As explained in the nearby panel, the current European specifications generally require a significantly higher performance from the oil than the US specifications.

Table 5.4 is complex in its own right but, on top of it, there are the specific in-house tests that some car manufacturers also require. More often than not, oils may have to meet specific manufacturers' performance specifications as well as the ACEA ones.

5.1.3 Japan

Historically, the manufacturers of Japanese cars have specified engine lubricants quite differently for their home market than they have for the same engines in export markets. In the home market they have tried to define and control lubricants in the same way as they do spare parts. They would like the owner of the car to use oils that are branded in the manufacturer's name. Often no performance specification is issued.

Obviously, such an approach was not suitable for markets outside of Japan, particularly if you were trying to build car sales in those markets from a low base. In non-Japanese markets, oils were specified in the terms that the local market was used to, and, at least in the early days, no attempt was made to tie consumers to the manufacturer's branded products. As the Japanese share of these markets has grown, they have introduced 'genuine' oils to many of them also. However, the Japanese manufacturers have never had the domination of lubricant sales in international markets that they have in their home market, in part because legislation exists in some of these other markets to prevent potentially-restrictive trade practices.

In recent years, the Japanese manufacturers have participated actively in the introduction of general performance specifications. In the early 1990s, the Japan Automobile Manufacturers Association (JAMA) joined with the American Automobile

Engine lubrication

Manufacturers Association (AAMA) to form the International Lubricant Standardisation and Approval Association (ILSAC) and develop oil specifications for gasoline engines in the US market. This is detailed in Section 5.1.1. They were forced to do this to have some influence on the oils generally available in the US market and, via this, the rest of the world. Many of their cars were sold, and in some cases made, in these international markets.

5.2 Viscosity level

As we've seen, up until the late 1980s, outside of Europe, the performance and viscosity specifications for engine oils in the general market did not seriously address viscosity stability. They specified only the viscosity of the fresh oil, and only under low shear conditions.

Multigrade oils – oils that contain viscosity index improvers – were introduced commercially in about 1964. Other than the requirement for the oils to pass the relevant engine performance tests – which certainly gave some assurance – there was no requirement for their viscosities to be maintained in service. This changed to a degree in 1989 with the introduction of SG and the requirement that oils must stay-in-grade during the ten-hour, so-called L38 engine test. This was not a particularly severe test but it was a significant change.

SAE viscosity grade	Low temperatures		High temperatures		
	Cranking viscosity (cP max at °C)	Pumping viscosity (cP max at °C)	Viscosity (cSt min) at 100°C	Viscosity (cSt max) at 100°C	High shear viscosity (cP min at 150°C)
0W	6200 at -35	60,000 at -40	3.8	-	-
5W	6600 at -30	60,000 at -35	3.8	-	-
10W	7000 at -25	60,000 at -30	4.1	-	-
15W	7000 at -20	60,000 at -25	5.6	-	-
20W	9500 at -15	60,000 at -20	5.6	-	-
25W	13000 at -10	60,000 at -15	9.3	-	-
20	-	-	5.6	<9.3	2.6
30	-	-	9.3	<12.5	2.9
40	-	-	12.5	<16.3	2.9*
40	-	-	12.5	<16.3	3.7**
50	-	-	16.3	<21.9	3.7
60	-	-	21.9	<26.1	3.7

*0W/40, 5W/40 and 10W/40 grades only

** all other 40 grades

Table 5.6 J300, SAE viscosity grades for engine oils.

Which Oil?

A much greater change followed in the early 1990s with the introduction of a high shear, high temperature (150°C) test requirement in the SAE J300 engine oil viscosity specification system (see the last column in Table 5.6). From that time on, these requirements have been incorporated into the AIP engine performance specifications (in SH, SJ, etc). In Europe, these HTHS requirements had been incorporated into the centralised oil performance specifications (G1, G2, etc) since the late 1970s. European motorists were much better served by the oil industry in this regard than their counterparts in the US.

During the same period (late 1980s to early 1990s), a great deal of work was also done on the performance of engine oils at very low temperatures. The mini rotary viscometer (MRV) and cold crank simulator (CCS) tests and specifications were introduced (see the left-hand columns in Table 5.6). They were followed, later in the 1990s, by the addition of a low temperature gelation test – a more practical replacement for specifications based on Pour Point.

The outcome is that, today, provided that the products are true to their labels, multigrade engine oils are far more shear stable than often was the case prior to 1990. This is true for both the US and Europe, but particularly for the US which suffered in the past from low shear stability products. The situation has in general been helped by the widespread introduction of synthetic or partially-synthetic engine oils. Many of these meet multigrade viscosity specifications without the need for the addition of any polymeric material. Hence they are completely shear stable. However, as explained in Section 4.1.2, not all synthetics have high VIs, and so you still need to check with the marketer/manufacturer of the particular oil you are considering.

5.3 Summary

In summary, since the 1950s or so, manufacturers have specified two properties in order to define those oils that are suitable for their engines. The first is the level of performance with respect to the oil's ability to control engine wear, cleanliness, corrosion, etc. The second is the appropriate oil viscosity to use, often expressed in relation to the anticipated ambient temperature in which the engine will operate. They also specified the appropriate oil drain interval.

In the US, the developed parts of Asia (with the exception of Japan) and in South Africa, Australia and New Zealand, the specification of required performance level has usually been done by reference to the API performance classifications, augmented recently in the US by the broadly similar ILSAC classifications.

In Europe, probably because of the number of countries involved, the car manufacturers and the oil industry were slower to introduce or adopt universal performance specifications. They tended to specify particular branded products for longer than was the case in the US. After a brief period in the 1960s and 1970s where the US specifications were used to some degree, the Europeans developed their own. This was, in general, a positive move for European motorists because the engine tests that had to be passed were (and still are) more severe and more extensive than those in the US specifications. The downside is that, unfortunately, the European specifications have tended to become more and more complex over time.

6. Transmission lubrication (including final drive)

For the purposes of this chapter I define automotive transmissions more broadly than is usual; I include the final drive. My reason for doing this is simply to avoid duplication, given that much of the discussion is applicable to all of the components involved. So, to be clear, by transmissions I mean the entire series of mechanical units provided to transmit the mechanical output of the engine of your car to its road wheels.

Under this definition, motor car transmissions usually consist of either a gearbox followed by a differential, or of a gearbox and transaxle, the latter two devices sometimes being contained in a single unit and sometimes being separate units. Because of the torque and power characteristics of internal combustion engines, it is usually a requirement that, as well as transmitting the engine power, changing its direction and splitting it to the driven wheels, the transmission must also assist the engine to operate in its optimum speed range, irrespective of the vehicle (that is, the road wheel) speed. This is achieved by means of sets of gears of different, progressive ratios, that can be changed while the vehicle is in motion. Also, one of the gear sets is designed to allow the vehicle to be driven backwards. The mechanism provided to allow the change between the gear sets may be manual or automatic.

The main requirements of the lubricants in the transmission are to:

- Control the wear of the gear teeth
- Control the wear of all bearings that support the shafts and other components
- Prevent rust or corrosion of the components
- Keep components cool
- Keep components clean

It is also necessary that the lubricant does not froth or aerate excessively, that it facilitates gear changes rather than hinders them, that it not react chemically with any of the components (including any seals), and that it not change its properties significantly in service (eg through chemical change, such as oxidation, or physical change, such as shearing, leading to a loss of viscosity).

Although the requirement list is a long one, overall the job of a gear oil is much simpler than that of an engine lubricant. Hence, transmission units usually do not have a pressurised lubrication system but instead rely on splash and dip. The exception is some automatic transmissions, but, even with these, the pressurising of the lubricant is usually to assist with the gear changing and torque conversion rather than the lubrication. In automatic transmissions the lubricant must also give satisfactory gearshift characteristics.

Which Oil?

6.1 Controlling the wear of the gear teeth

The degree of difficulty of this task depends on the design of the gear teeth and the loads that they have to transmit. Ideally, full-film lubrication should be maintained between the meshing teeth. As has been explained (for example, in Section 3.5), the higher the viscosity of the lubricant the greater the load range where full-film lubrication can be maintained. However, in a typical transmission, the oil is thrashed around by the rotating components and, if the viscosity is too high, power will be lost and the oil will get hot. Again, there's an optimum viscosity.

If the load is sufficiently high, or the geometry is difficult, then full-film lubrication will not be able to be maintained. Mixed film or boundary lubrication may prevail, and the oil will have to contain some form of anti-wear or EP additive to prevent excessive wear. The loads are set by the design of the transmission and the amount of power that it has to transmit (the specification of the engine). The geometry is also set by the design. Thus, the car manufacturer must specify both the required viscosity of the lubricant and the required level of ability (if any) to handle non-full-film lubrication.

6.1.1 Viscosity

Starting with viscosity, for many years now the viscosity of lubricating oils for motor

SAE viscosity grade	Viscosity cSt at 100oC (min)	Viscosity cSt at 100°C (max)	Max T for viscosity of 150 poise (°C)
70W (new)*	4.1	-	-55
75W (old)*	4.1	-	-40
75W (new)	4.1	-	-40
80W (old)	7.0	-	-26
80W (new)	7.0	-	-26
85W (old)	11.0	-	-12
85W (new)	11.0	-	-12
80 (new)	7.0	Less than 11.0	-
85 (new)	11.0	Less than 13.5	-
90 (old)	13.5	Less than 24.0	-
90 (new)	13.5	Less than 18.5	-
110 (new)	18.5	Less than 24.0	-
140 (old)	24.0	Less than 41.0	-
140 (new)	24.0	Less than 32.5	-
190 (new)	32.5	Less than 41.0	-
250 (old)	41.0	-	-
250 (new)	41.0	-	-

* Old refers to pre-2005, new to 2005 and beyond.

Table 6.1 SAE J306 viscosity grades for automotive gear oils.

Transmission lubrication (including final drive)

vehicle transmissions has been specified almost universally by means of the SAE system (see Section 3.1). For automotive gear oils the specification is called SAE J306. You should note that it is not the same as the similar SAE system for engine oil viscosity (SAE J300). The system can be traced all the way back to 1923, but was not presented in its present form until the 1930s. It then remained essentially unchanged up until 2005, and is summarised in Table 6.1.

Although the viscosity classification system remained basically the same for over 60 years, it unfortunately contained a major flaw. The viscosity grades it specified were very wide. This problem was finally addressed in 2005 when the system was revised. The new (and current) version is also summarised within Table 6.1. To simplify the comparison, the old and new viscosities are also plotted in Figure 6.1.

The revision (actually two sequential revisions) introduced:

- A new, lower viscosity winter grade (SAE 70W)
- Four new high temperature grades (SAE 80, 85, 110 and 190)
- A narrowing (and potential lowering) of the high temperature viscosity of the SAE 90 and 140 grades

Most of the cars of interest to us were manufactured during the period when the old system applied, or even prior to its development. The car manufacturers' requirements were, therefore, probably specified under that system. So, today we have the challenge of not only understanding what the manufacturer of our car was specifying at the time that it was built, but also of translating that understanding across to today's oils, under the potentially confusing situation where, although much of the terminology used then and now is the same, its meaning is potentially slightly different.

Although gear oils and engine oils use different SAE numbering systems (compare Tables 5.6 and 6.1), the viscosities that the two systems specify overlap. This is shown in Figure 6.1. You will see that, at 100°C, the viscosity of an SAE 50 engine oil equates to the viscosity of an SAE 90 gear oil (old scale). You will also note that, because of the wide gear oil viscosity grades under the old system, an SAE 40 and an SAE 60 engine oil may also equate to an SAE 90 gear oil. Make sure that you understand which oil type is being referred to when a manufacturer specifies the use of a particular SAE viscosity grade.

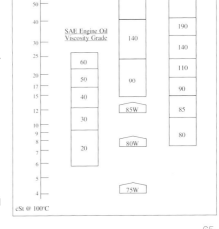

Figure 6.1

Which Oil?

> **Source of SAE gear oil viscosity grade numbers**
>
> If you're wondering where the gear oil viscosity grade numbers – the 90, 140, etc – came from I can give you a clue. The approximate midpoint viscosities of the three grades in the pre-2005 system, when measured at 210°F and expressed in units of Saybolt Universal Seconds, were:
>
> | | |
> |---|---|
> | SAE 90 | 90 SSU |
> | SAE 140 | 150 SSU |
> | SAE 250 | 250 SSU |
>
> The SAE 250 grade is actually open ended but 250SSU is a typical viscosity (at 210°F).

6.1.2 Mixed lubrication

Turning to the need to handle mixed film or boundary lubrication, it took longer for the combined automotive/oil industries to develop a consistent and effective system of categorization of gear lubricants from this perspective. Historically, the situation was similar to that which existed with engine oils – gear oils were classed as mild, regular, heavy or multipurpose – but the bases for these classifications were relatively vague. During World War 2 the US army took the initiative and developed a performance specification for those gear oils suitable for use in hypoid gear systems (boundary lubrication conditions prevail in these systems). It was based on mechanical testing, not on the chemistry of the oils, and thus actually predated the adoption of this approach for the specification of the performance of engine oils.

In 1950, this military specification was renamed MIL-L-2105 and it began to be used for non-military applications. The US oil industry responded by developing, under the auspices of the American Petroleum Institute (API), a system which is summarised in Table 6.2. It too was (and is) based on mechanical test performance (other than for the lowest performance level) and it covered the full gamut of gear lubrication demands by defining six classifications or performance levels (GL-1 to 6). The MT-1 classification was added quite recently, and is mainly relevant to certain manual gearboxes used in trucks.

As a broad generalisation, the gear teeth in gearboxes in motor cars do not experience boundary lubrication in normal operation, and do not require full EP additive protection. However they very commonly experience mixed film conditions, hence the normal requirement is for an oil of GL-3 (or sometimes GL-4) performance level, an oil with an appropriate level of anti-wear additive present.

It's also generally true that the gear teeth in hypoid differentials (Figure 6.2) do experience boundary lubrication. The common lubrication requirement for these is an oil with a GL-5 performance level – an oil with full EP additive treatment present.

Transaxles vary in design, but most often they require only GL-3 (or sometimes GL-4) performance. Usually, they do not contain offset gears and – depending on loads – the gear teeth do not experience boundary lubrication. However, there are exceptions.

Transmission lubrication (including final drive)

API classification	Type or description	Typical applications
GL-1	Straight mineral oil.	Certain truck manual transmissions.
GL-2	Usually contains fatty materials.	Worm gear drives, industrial gears.
GL-3	Contains mild EP (antiwear) additives.	Manual transmissions and spiral bevel final drives.
GL-4	Equiv. to obsolete MIL-L-2105, usually satisfied with 50% GL-5 additive level.	Manual transmissions, and spiral bevel and hypoid gears in moderate service and geometry.
GL-5	Virtually equivalent to latest* military specification, typical recommendation for most car and truck differentials.	Moderate and severe service in hypoid and other types of gears. May also be used in some manual transmissions.
GL-6	Obsolete.	Severe service involving high-offset hypoid gears.
MT-1	Contains thermal stability and EP additives.	Non-synchronised manual transmissions in heavy duty service.

* Currently SAE 2360. Was MIL-L-2105D, then MIL-PRF-2105E.

Table 6.2 API gear oil service designations.

If your car has a manual gearbox, it may not be immediately obvious – when you look up what its manufacturer specified – just what level of load or EP performance was required from its lubricant. For many cars of the 1950s and 1960s, the manufacturers specified monograde engine oil for this application. However, the top line engine oils in that period all contained zinc dithiophosphate (ZDTP) anti-wear additive in order to handle the lubricant requirements of the cams and followers. They were, in fact, of GL-3 performance, even though this was seldom if ever stated.

You may be tempted to think that, if anti-wear (GL-3) performance is necessary for my gearbox, then using an oil that can handle even higher loads – one of GL-5 performance with full EP additives – would be even better for it. Resist this temptation. To give the EP performance, most GL-5 lubricants use compounds that contain sulphur, and sulphur is chemically very aggressive to what are sometimes called the 'yellow metals.' These nonferrous metals include copper and alloys of copper, such as bronze. Many gearboxes use such metals for bushes or in synchronisers and, at operational temperatures, they can be attacked by the EP additive.

The visual appearance of such chemical attack can vary from glazing to

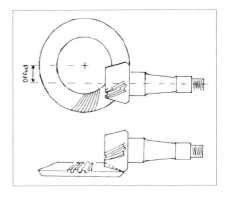

Figure 6.2

Which Oil?

corrosion or pitting of the metals concerned. In the 1980s I drove a 1969 Fiat 124 AC Sport for my daily commute. Fiat specified an SAE 40 engine oil for its manual gearbox. By the 1980s, monograde engine oils had become hard to find in service or gas stations, and so I asked a local workshop that specialised in Fiats, Lancias and Alfas what it used. The mechanic said that he used a particular brand of diff oil. When I queried this, he stated that he had never had a problem. I then asked whether or not the synchronisers tended to pit. "Oh yes," he said, "they all pit!" The mechanic viewed the pitting that he observed as a 'feature' of the Fiat gearbox, not a consequence of his choice of lubricant.

6.1.3 Use of other oils

It is possible today to formulate oils that have EP performance and that, while based on sulphur or other additive chemicals, do not attack copper aggressively. You may find oils marketed as being of full EP (GL-5) performance and with the statement that they are suitable for use in manual transmissions. However, I would always err on the side of caution. Check with the lubricant manufacturer (not the marketer) before using it in an application where the original manufacturer of the gearbox did not specify an EP level of performance.

Another temptation that you should resist is to use a conventional multigrade engine oil in your transmission. At the time (1980s) that I had the predicament with my Fiat's gearbox and could not obtain an SAE 40 monograde engine oil, forecourts were littered with multigrades, such as SAE 15W/40 or 20W/40. These were of the correct viscosity at 100°C, and they all contained ZDTP anti-wear, so why did I not simply use one of them?

The reason is that the polymers used in these conventional, mineral oil-based engine oils to give their multigrade viscosities were (and still are) not anywhere near shear stable enough to survive the threshing that they would receive in a gearbox. These oils will rapidly and permanently drop in viscosity. If you should find yourself in a situation where you have no alternative other than to use a conventional multigrade engine oil to get you out of trouble, then use the one that is of the narrowest multigrade range that you can source – for example, use an SAE 20W/40 rather than a 15W/40. It will have the lowest polymer content, the least potential shear loss, and the highest inherent viscosity of its base oil. Ensure that you change the oil back to the correct oil as soon as you're able.

The above warning does not apply to fully-synthetic multigrade oils (provided they contain appropriate levels of anti-wear additives). They generally do not contain polymers which are vulnerable to shear, and they have inherent multigrade characteristics.

As the 1980s progressed, manufacturers

Property	Typical gear oil (SAE90)	Typical ATF
Viscosity, cSt@40°C	150	35.5
Viscosity, cSt@100°C	15	7.2
Viscosity Index	97	172

Table 6.3

Transmission lubrication (including final drive)

of manual gearboxes woke up to the fact that monograde engine oils were no longer readily available. They looked around for a convenient alternative that had the required anti-wear performance. Many of them came up with Automatic Transmission Fluid (ATF – see Section 6.2 following). These oils are designed to handle the lubrication of the gears in an auto box (as well as the wet clutches, bands and torque converter), and they contain anti-wear additive. However, ATFs have much lower viscosities than the oils that are typically specified for manual transmissions (see Table 6.3).

The disappearance of monograde oils coincided with the rapid move in conventional car design from rear-wheel drive to front-wheel drive. Long and complex linkages were now needed from the shift or gear lever to the gearbox. The use of the lower-viscosity ATFs in these gearboxes had an advantage. They helped the 'shift feel' at low temperatures. In this context, 'poor shift feel' was often a euphemism for 'can't get it into gear!'

If you lived in a cooler climate, then this solution was generally fine. However, if you lived in a warmer part of the world, the use of ATFs often gave shifter or gearstick rattle when the gearbox reached operating temperature, and, in some cases, gear wear. **Do not** use an ATF in an older manual gearbox that did not specify it, other than in an emergency.

Fortunately, many lubricant marketers (and car manufacturers) have now addressed the loss of monograde engine oils, and the less-than-ideal performance of ATFs. Today, there are many GL-3/GL-4 performance oils sold which are designed specifically for use in manual transmissions. The problem now is not so much to find a product, but to choose the best and most appropriate one from the range of offerings.

6.1.4 Multigrade transmission oils

Many of today's gear oils are genuine viscosity multigrades. In recent years some very highly stable polymers have been developed that can withstand shear far better than the ones used in engine oils. Also, synthetic oils that change their viscosity with temperature much less than mineral oils do, have been developed. These synthetics may meet the multigrade viscosity specifications in their own right, without the necessity to add any VI improver polymer, and they are completely shear stable. So, fortunately, it should now be possible to choose an oil for your manual transmission that will have the appropriate viscosity when hot, but also have good shift feel and synchroniser performance when cold.

In the previous paragraph I used the phrase 'genuine multigrades.' Back in the 1930s, when the oil and motor industries standardised the definition of the viscosity grades of gear oils, they gave them very wide viscosity ranges (see Figure 6.1). This was understandable in an era that was just emerging from definitions such as 'heavy' or 'light.' However, in the 1960s, the oil industry exploited the width of the viscosity grade definitions when it introduced multigrade gear oils which, unlike the situation with engine oils, did not require the use of any VI improver polymer. They were formulated on straight base oil (plus appropriate load-carrying additives) but,

Which Oil?

because the refining of mineral base oils had improved, the inherent VI of the base stocks had increased and the wax content had decreased.

The viscosity of these newer base oils changed less with temperature than had been the case with the earlier, less refined ones. The point was reached where, if a gear oil were formulated to have a viscosity at the very bottom of, say, the SAE 90 scale, it now also met the SAE 80W specification (Table 6.1). Suddenly, commencing in the US and spreading to Europe, SAE 80W/90 gear oils (and SAE 85W/140 diff oils) became the norm. They replaced SAE 90 (and SAE 140) oils. There was no false representation or claim involved. The oils did meet both viscosity grades but, unlike engine oils (for which the public had developed some understanding), no viscosity modifying additives were used or necessary.

However, the viscosities of the oils, as formulated and sold, did drop significantly in comparison to what had been on the market previously, even though the stated grades did not apparently change. The previous 'straight' SAE 90 or 140 products had been formulated at the middle of the specified viscosity ranges.

That this drop or reduction was of some potential concern to the oil companies is shown by the fact that, for heavy industrial use, they all continued to market straight SAE 90 (and SAE 140) grades. These products had the same viscosity indices and wax contents as their automotive cousins – being formulated on the same base oils – but their viscosities were pitched in the middle of the respective SAE ranges, rather than at the very bottom. Even though they had the same viscosity/temperature characteristics they could not claim to be multigrades.

With the recent development of highly shear stable VI improvers, and also the introduction of synthetic base stocks, it has become possible to produce, say, an SAE 80W/90 gear oil where the 100°C viscosity is right in the middle of the old SAE 90 range (not at the very bottom) and the product also meets the SAE 80W specification. Such an oil would be the equivalent (at 100°C) of the 'industrial' gear oil version, but would be what I consider to be a 'true' multigrade. It was the potential to have two oils with the same SAE viscosity claim, but significantly different actual viscosities, that forced the recent upgrade and tightening of the definitions. However, be aware that the new definitions of both the SAE 90 and the SAE 140 viscosity grades – historically the most commonly specified gear oil grades – now lie in the bottom half of the old ranges (Figure 6.1). Accordingly, to produce an SAE 80W/90 gear oil under the new scale, it is still not necessary to add a VI improver. An SAE 80W/110 oil would require the use of VI improver, or at least some synthetic base stock.

Faced with a choice between two EP diff oils – one branded as an SAE 90 and the other as an SAE 80W/90 – under a circumstance where the manufacturer of your 1956 vehicle specified an SAE 90, what should you do? If you were making this decision prior to 2005, you could have presumed that both oils would have had similar viscosity vs temperature characteristics, but that the SAE 90 would have been a heavier oil than the SAE 80W/90. You should have chosen the SAE 90 product because that is the type of product that existed in 1956.

Post-2005, you would presume that both products have essentially the

Transmission lubrication (including final drive)

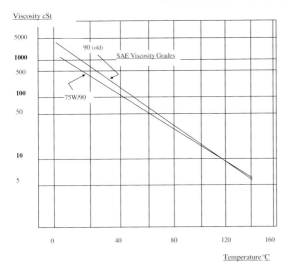

Figure 6.3

same viscosity (and viscosity/temperature characteristics) and that, for both of them, the viscosity at 100°C is lower than was the case in 1956. It does not matter which you choose. However, you should consider using an SAE 75W/90 product. Although this will also be lower in viscosity at 100°C than the 1956 product, it will be as thick as the old product at higher temperatures, and will do a better job at low temperatures (see Figure 6.3). However, I am now encroaching on Chapter 8's territory, and you should go there for further guidance.

6.1.5 Limited-slip characteristics

Some higher performance cars use what are called limited-slip differentials. Modern versions may be based on viscous couplings, but here I am talking about the older, mechanical or torque-sensitive type which contain wet clutches. To ensure correct operation, and to control wear over time, it is important that the lubricant used in these differentials has the correct viscosity and friction characteristics. Use an oil that is specifically marketed for limited-slip use. It will contain a friction modifier additive. Other than this aspect, all other lubricant requirements of these differentials are the same as those for the standard versions.

6.1.6 Overdrives

Under this heading I deal only with the 'classic' style of overdrive, the 'bolt on' units such as those manufactured by Laycock de Normanville (GKN) and Borg-Warner. The overdrive typically consists of an electrically- or hydraulically-operated epicyclic gear train located between the gearbox and the differential on rear-wheel drive vehicles. When engaged, it gives a fixed drop down in the rotation rate between these two units, effectively shifting up in gear. Engagement is usually manual in the Laycock unit, and semi-automatic in the Borg-Warner.

Overdrive units generally contained bronze components, and so their manufacturers recommended against the use of oils that contained EP additives (hypoid gear oils). Although they did not use the terminology, they required a GL-3 gear lubrication performance level. The usual recommendations were SAE 30 or

Which Oil?

40 engine oils, or non-EP SAE 80 or 90 gear oils. Multigrade viscosity oils were not recommended – shear of the polymeric viscosity index improvers would have been a potential problem.

In some applications non-friction-modified ATF was recommended (eg Ford Type F). This was usually to simplify oil inventories. However, the viscosity of the ATF was lower than that of the recommended engine oils and, in some cases, the recommendation was later changed back to engine oil.

6.2 Automatic transmissions

The designs of automatic transmissions vary, and, in recent years, there has been a trend – particularly in top end and higher performance cars – to introduce automatic or powered dual clutches, coupled to an essentially conventional manual gearbox with a powered shift. In this book I am dealing with the 'traditional' automatic transmission. This usually consists of a gearbox with planetary gears, a set of wet clutches and bands to bring the appropriate gears into play, and a torque converter to allow the car to sit stationary with the engine in operation (and also to multiply the torque). All the components are located within a single housing and use a common lubricant.

As well as the characteristics listed at the start of this chapter for transmission oils in general, the lubricant used in an automatic transmission must also have (and must retain in service) desirable friction characteristics. This is to allow gear changes to occur efficiently and smoothly.

The US led the way with automatic transmissions for mass-produced cars. The result was that the lubricants specified by the major American car manufacturers for their automatic transmissions became pseudo world standards. Unlike the situation with engine and gear lubricants, standards were not set by industry bodies, either automotive or oil.

Of the major US manufacturers, it was GM who pioneered the system that we know today. It introduced a semi-automatic transmission in 1937 in its Oldsmobile Division, but it didn't catch on. However, it was a different story when, in 1939, Olds introduced a modified version under the name Hydra-Matic. In this version, a fluid

---

Japanese automatics

In general, with their smaller engines and lower targeted price range, all early Japanese cars had manual transmissions. As the Japanese manufacturers' market share and spread increased in the late 1960s, into the 1970s, they introduced automatic transmissions, usually via the route of using specialist US transmission companies or licensing US manufacturers' technology.

A major exception was Honda. Faced with developing an automatic transmission that was compatible with its then tiny 360cc or so engines, it could not find any company that had relevant experience. Using a Borg-Warner box as a base, it developed its own technology.

In general, all Japanese automatics in older cars use a friction-modified fluid. The DEXRON family is suitable.

Transmission lubrication (including final drive)

coupling replaced the mechanical clutch, and the operation was entirely automatic (other than the selection of reverse). It was a marketing (and technical) sensation. Ford followed in 1950 with the Ford-O-Matic, a transmission developed by Borg-Warner, and Chrysler in 1954 with its Powerflite. While some of these early units had fluid couplings rather than torque converters, the principle of operation was essentially the same as the 'traditional' automatic transmission that we know today.

The first automatic transmissions used engine oil for the lubricant/hydraulic fluid, with the recommendation to add some sulphurised sperm whale oil supplement. The oil was dyed red to make it easier to see leaks and distinguish them from engine oil. The red colour is the one aspect of the lubricant that has not changed in 60 years.

Partly because the addition of the sperm whale oil supplement in service was something of a hit-or-miss affair, and partly because the torque converter that had replaced the fluid coupling initially used generated more heat, GM introduced a special lubricant in 1949. It contained an additive to modify (reduce) friction, and a higher level of antioxidants. The idea of the friction modifier was to lengthen the gear change time, to make it smoother and reduce wear. The lubricant (or fluid as these lubricants tend to be called) was designated as Type A by GM, and then – when its oxidation performance was upgraded further in 1957 – as Type A Suffix A.

Perhaps just to be different from its great rival, Ford decided to abandon friction modification, and so, in 1961, it introduced a non-friction-modified fluid under the Specification ESW-M2C33-D. This specification evolved fairly rapidly so that, by 1967, the then-current version was ESW-M2C33-F; the fluid that has become known universally as Type F, probably because people imagine that the letter F stands for Ford.

The General Motors Type A Suffix A fluid became the pseudo world standard. This was in part because a number of European manufacturers introduced GM-sourced automatic gearboxes into their own cars. The fluid was upgraded in 1967 and renamed DEXRON, but it was still friction-modified.

In the early 1970s, just after a further upgrade to DEXRON II, GM ran into a problem. It changed the alloy composition of the solder used on the auto transmission oil coolers – probably to cheapen it – and some customers experienced corrosion problems in the field. Rather than upgrading the solder, GM required the ATF to now have a corrosion inhibitor added to it. The modified fluid was designated DEXRON IID and, in 1975, it replaced the already 'new' DEXRON II, which became known retrospectively as DEXRON IIC.

However, there was an unfortunate side effect. The presence of the corrosion inhibitor – which was polar in nature – rendered the new fluid much poorer at shedding water than the old one. This was not a particular problem in the automotive area (which was GM's concern) but it was in other areas where the fluid had become the standard product – areas such as hydraulics.

Many non-GM companies now had a quandary. They did not want to move their specification to DEXRON IID, but to keep it as DEXRON could be inviting problems. Users would not necessarily distinguish between the two products, particular as

Which Oil?

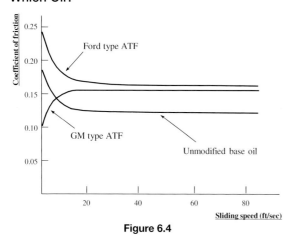

Figure 6.4

GM itself stated quite uncategorically that you could use the new DEXRON IID in all of its earlier automatic transmissions – back to the first – and it was right.

The outcome was that a number of manufacturers reverted to specifying Type A Suffix A fluid, one of the more conspicuous being Mercedes-Benz. It's unfortunate that this confusion arose because, in reality, for automotive use, DEXRON IID was entirely suitable for all transmissions that required a fluid with friction modification, and this was essentially every transmission other than those in Ford vehicles or, possibly, those of Borg-Warner derivation.

Other than this hiccup, ATF technology and specifications stayed fairly stable through most of the 1960s and 1970s (Table 6.4). However, the drive for fuel economy was catching up. Devices such as lock-up torque converters were developed, and they were prone to juddering. For these, Ford was forced to use friction modification, and they it introduced Type CJ fluid in 1974. The non-friction-modified Type F also morphed into Type G in Europe, and then rapidly into Type H, which actually had some friction modification.

The confusion in the Ford customer camp was now such that Ford abandoned the Type designation altogether and introduced a new name – MERCON – to try to tidy things up. All MERCON fluids are friction-modified and are not recommended for the older Ford transmissions that specified non-friction-modified fluids.

Over at GM, things were more straightforward. DEXRON II, albeit with a series of letter changes after its name (IIE etc), ran right through until 1993 when GM introduced DEXRON III. The introduction included the statement that it was backward compatible with all of their earlier transmissions – all the way back to 1949. Technically this is true but, in the interest of fuel economy and low temperature shift characteristics, DEXRON III has a lower viscosity at low

---

Friction modification

With all the discussion of friction-modified vs non-modified ATF fluids, you may wonder how significant the difference between them is. Figure 6.4 shows the variation of the friction coefficients of the two fluids in a sliding test. The Ford type of fluid has a characteristic similar to an unmodified mineral base oil. By contrast, the GM fluid has a quite different frictional behaviour as the sliding speed approaches zero.

Transmission lubrication (including final drive)

Year	Specification	Internal no	Company	Friction modified
1949	Type A		GM	✓
1957	Type A Suffix A		GM	✓
1959	M2C33-A		Ford	✓
1961	M2C33-D		Ford	✗
1964	MS-3256		Chrysler	✓
1966	MS-4228		Chrysler	✓
1967	DEXRON		GM	✓
1967	M2C33-F (Type F)			✗
1973	DEXRON II(C)	6137-M	GM	✓
1974	M2C138-CJ		Ford	✓
1974	M2C33-G (Type G)		Ford (Europe)	✗
1975	DEXRON IID	6137-M	GM	✓
1980, May	MS-7176 Change D		Chrysler	✓
1981, June	MSC166-H		Ford	✓
1987, Jan	MERCON		Ford	✓
1990, Oct	DEXRON IIE	6137-M	GM	✓
1992, Aug	DEXRON IIE Rev	6137-M	GM	✓
1993, April	DEXRON III	6297-M	GM	✓
1996, July	MERCON V		Ford	✓

Table 6.4 Approximate timeline of ATF specifications.

temperatures than the older fluids. It has a higher VI. Not every old GM transmission in use is oil tight, and DEXRON III will usually do an even better job than DEXRON II in finding a spot to leak through.

Automatic transmissions have continued to get more and more complex. Six speeds are now the norm, as are electronic controls. Also, the move to front-wheel drive has made their packaging and their cooling more difficult. The ATF specifications have continued to evolve at a pace. We now have DEXRON IV and are still counting, albeit in Roman numerals.

7. Chassis, steering and wheel bearing lubrication

The lubrication needs of these three areas – chassis, steering and wheel bearing – are different. The relative movements of surfaces in chassis components are usually quite small, and often oscillating. Movements in steering components are similar, albeit over a larger range. For bearings, there is a continuous, unidirectional movement while the vehicle is in motion. In spite of this, I have grouped their lubrication in a single chapter because, in general, the form of lubricant used in all three applications is a grease. I have further grouped steering and chassis under the one heading or phrase – chassis lubrication. This is because the lubricants specified for the two applications over the years have tended to be the same.

Greases are used to lubricate moving surfaces in those situations where an oil could not be retained; it would run out. Greases, being semi-solids, can stay in place. A modern grease – and by modern I merely mean one that has been developed in the past 100 years! – is an oil that has been thickened by the addition of an appropriate finely-dispersed filler. Typically it may be 90 per cent oil. The filler may be an inert substance such as clay, but, in most cases, it is a soap. The soap is specially made from fatty materials so as to have desirable properties.

The word 'grease' predates the development of modern greases by thousands of years. Many animal fats are liquids when warm but soft solids when cool. Such substances – in their cool state – were called greases. They have lubricating properties and usually are adhesive and water resistant. As a result, some of them were drafted into service as the early chassis and wheel bearing lubricants. Tallow was the most common substance used and it did a completely adequate job for many years, first in horse drawn vehicles and then in motor cars.

Elastohydrodynamic lubrication (EHD or EHL)

The lubrication of rolling element bearings, such as the typical ball or roller bearing used in hubs, is a very complex matter. It is not covered by the earlier discussion of lubricant films and their integrity.

In a ball bearing, although there is a rolling motion, there is a point contact. Loads are incredibly high at that tiny, localised point, and, if the film situation applied, the lubricant would be squeezed out.

This does not happen for two main reasons. The first is that, because the load is so high, the rolling element and the race actually distort physically at the point of contact. A point never occurs. Second, the viscosity of a liquid increases with the pressure applied to it. The pressures are so high that the liquid essentially becomes a solid as it passes through the contact area. It is not squeezed out but instead forms a protective film. Extreme pressure additives are not needed. This lubrication situation is called elastohydrodynamic lubrication – EHD or EHL.

Chassis, steering and wheel bearing lubrication

Of the two jobs – chassis and wheel bearing lubrication – wheel bearing lubrication is potentially the more complex. However, in the early days of motoring – in fact through until at least the 1930s – the demands on the wheel bearing or hub lubricant were fairly simple and straightforward. Speeds and loads were moderate, brake performance was modest (and often not on all four wheels), and operating temperatures were close to ambient. The duty was not severe.

Even so, if a source of oil were available – such as can be the case with the bearing(s) in a hub that is attached to a rigid driving axle with a differential inboard – it was common to use this oil as the lubricant for the bearing. However this approach necessitated that an oil seal be installed outboard of the bearing to prevent the oil being lost, both from the bearing and, even more critically, from the differential.

If no oil source were available – as was the case with a non-driven wheel or with a driven one which did not have a closed connection to an oil source (for example, where independent suspension was used) then grease was the form of lubricant that had to be used. As mentioned it has the advantage of staying in place and not running out of the bearing. Only very simple seals are needed – as much to keep the dirt and water out as the grease in.

The task for the chassis grease was even simpler. The main operational difficulty faced was water, either during rain or when crossing drains, creeks, etc. Thus, the grease had to have high water resistance – both resisting being washed out and also not emulsifying or dissolving in water. This characteristic was also fine for the bearings, and so the same grease could be used for both jobs.

7.1 Types of grease

Natural greases, such as beef tallow, are highly water resistant, highly adhesive, and good lubricants. Their main drawback is a limited range of operational temperature.

Molybdenum disulfide (Moly)

Molybdenum disulfide is a dark grey to black solid powder that is sometimes added to greases. It can be very beneficial in situations where the surfaces have to slide over each other and loads are very high. It is widely used in heavy-duty industrial greases.

In automotive applications, its presence in a grease can be beneficial in chassis lubrication. It is of no real benefit in rolling element wheel bearings, though, and can sometimes be detrimental. In continuous movement, the solid moly will tend to separate out into the regions of least shear. The result is a thickening of the grease in those areas, reducing its ability to flow back into the high shear areas of the bearings where the loads are the highest and lubrication is most needed. This problem does not generally arise in industrial applications because bearings are regularly re-greased. However, this is not the case in automotive bearings where there is usually a long period between grease replenishments.

Overall, moly is beneficial in the grease used for all chassis lubrication applications, but of little or no value for wheel bearings in most vehicles. A moly level of 3 per cent is adequate, and higher levels will not necessarily perform better.

Which Oil?

At low temperatures they become hard, and at high temperatures they become too fluid. However, for many years they sufficed.

Technology advanced and methods were developed to make what I will call tallow-like greases. A soap was made by reacting tallow with lime (a form of calcium), and mineral oil was then mixed with it. These greases are known as calcium greases and, while they share tallow's characteristic of excellent water resistance, they harden less at low temperatures. Calcium greases still have the weakness of not being able to handle high temperatures. At 90-100°C they lose their structure and become fluid. Their maximum safe operating temperature is about 80°C.

80°C is an adequate maximum operating temperature for a chassis grease, and for many years it was also satisfactory for a wheel bearing grease. However, as braking performance improved (and cars also became faster and heavier) the amount of heat generated within the brakes increased. This is because, during braking, the forward kinetic energy of the car is converted to heat at the friction surfaces of the brakes. These surfaces were (and still are) usually located within or adjacent to the wheel hubs, and so the wheel bearings also became hot.

The technical response was to develop greases with higher melting points. These greases were introduced commercially in the 1940s and typically were based on sodium soaps. Sodium-based greases have a melting point of the order of 150-160°C, and an operating range up to about 120°C. Their appearance is quite different to the very smooth-textured calcium greases. They have a fibrous appearance, and were made in both short and long fibre forms. The short fibre versions were more shear stable, and thus were preferred for use in bearings.

While they solved the melting problem in wheel bearings, sodium greases, unfortunately, do not have good water resistance. They could not be used as chassis greases, and so, as the 1940s and '50s progressed, car manufacturers increasingly had to specify the use of two different greases.

This annoyance was solved by technical advances made during the Second World War. Greases based on lithium soaps were developed. Although more expensive than both calcium and sodium greases, they had a higher melting point than a sodium grease (180 vs 160°C), and also had excellent water resistance and good shear stability. After some delay, by the early 1960s lithium soap greases became widely available. They were called multipurpose greases, and they became the specified grease for both chassis and wheel bearings. Simplicity had returned.

Alas, it was not to last. Disc brakes were also developed in the 1960s and soon

---

Mixing different greases

Unlike the situation that applies with engine oils (provided they are designed for the particular type of application), greases cannot necessarily be mixed safely. What is critical is not what type of use the grease is designed for – bearing, chassis, etc – but what type of filler it uses. Some fillers (thickeners) will collapse if mixed with a grease that contains a filler of a different chemical type, and the grease will lose its structure. Table 7.1 gives some guidance but, in general, avoid mixing different chemistries. Brand or function are not the issue, chemistry is.

Chassis, steering and wheel bearing lubrication

	Lithium	Lithium complex	Calcium	Sodium	Clay (Bentone)	Polyurea
Lithium	✓	✓	✓	✗	✗	✗
Lithium Complex	✓	✓	✓	✗	✗	✗
Calcium	✓	✓	✓	✗	✗	✗
Sodium		✗	✗	✓	✗	✗
Clay (Bentone)	✗	✗	✗	✗	✓	✗
Polyurea	✗	✗	✗	✗	✗	✓

Table 7.1 Compatability of different greases.

became the norm – at least for the front brakes – initially in Europe, and later in the US and the rest of the world. The greater braking performance resulted in still higher hub and bearing temperatures, and a 180°C melting point for the grease lubricant was no longer adequate. A number of greases were developed with higher melting points. The most widely used initially were clay greases. These use a special clay powder as the thickener, and have no melting point. However, they are very sensitive to contamination with other greases. They were followed by the so-called complex greases (usually lithium complex or calcium complex). These have a melting point of over 250°C and, since 1990, have been widely adopted. More recently, polyurea greases have appeared.

All of these greases (clay, complex, polyurea) are more expensive than a straight lithium grease, but they can be used in both chassis and bearing applications, having good water and shear resistance, as well as high temperature capability. However, the reality is that, by the 1990s, chassis lubrication had become largely academic for the owner of a motor car. Grease nipples had disappeared, and all chassis components that required grease lubrication were sealed for life. In service, they had to operate with only the lubricant that had been added at the time of manufacture. When wear occurred, the entire component had to be replaced, again with a sealed-for-life unit.

7.2 Consistency of grease

Quite independent of the chemistry used to make the grease – calcium, lithium, etc – the other important property from a service perspective is the consistency or 'thickness' of the grease. Greases can range from semi-fluids to semi-solids in their physical consistency, and so it's important to be able to specify this property.

For many years, this has been done by the universal use of the National Lubricating Grease Institute (NLGI) Grade or Number system. The consistency of a grease is measured by letting a standard metal cone sink into it for a defined period of time (five seconds). The distance that the cone sinks, measured in tenths of millimetres, defines its consistency grade (see Table 7.2). For automotive use, greases of NLGI Consistency Grade 2 are almost invariably specified. One exception can be steering boxes where a Grade 1 grease – a sloppy grease – may be required. Thicker greases (Grade 3) may be specified for use between the leaves of springs or

Which Oil?

NLGI grade	Penetration (mm/10)	Texture
00	400-430	Thickened oil
0	355-385	Soft, semi-fluid
1	310-340	Soft, slumpable
2	265-295	Medium
3	220-250	Firm
4	175-205	Hard

Table 7.2 NLGI grease consistency grades

in spring shackles, and a semi-solid Grade 4 may be specified for water pumps. However, No 2 greases are by far the most commonly required.

7.3 Grease performance specifications

Historically, the specification of appropriate grease lubricants by car manufacturers has not been on the basis of reference to a performance specification. Manufacturers have either specified particular branded products or, more often, simply the thickener type and consistency grade. For example, they might specify a multipurpose lithium grease of No 2 consistency.

There is a very simple explanation for this. Until quite recently, there were no generally-agreed performance specifications for greases for automotive use. In an attempt to correct this situation, in 1990 the US grease manufacturing industry – through its industry body the National Lubricating Grease Institute (NLGI) – developed two sets of specifications. They relate respectively to greases for automotive chassis use (designated by the letter L) and greases for automotive bearing use (the letter G).

There is a hierarchy of performance classifications within these two groups – LA, LB and GA, GB and GC – with the level of performance increasing with the sequence (for example, GC is higher than GA). The NLGI also runs a formal certification system to ensure that claims made by grease marketers are accurate. However, it will only certify greases that meet the highest performance levels (LB or GC).

The NLGI may disagree with me but the two specifications have not swept the world. There is certainly nothing wrong with them. The general lack of knowledge and interest amongst mechanics and drivers is probably just a reflection of the fact that their development came too late. The need for chassis lubrication in service had already largely disappeared by 1990, when the specifications were introduced, and the re-greasing of wheel bearings was also relatively rare. The need for the specifications had already largely passed.

For we owners of older cars, their manufacturers obviously did not use these specifications to specify the appropriate grease or greases to use on their vehicles. The specifications did not, as yet, exist. However, they are appropriate specifications for our vehicles, and you can safely, and with potential benefit, use greases carrying the appropriate NLGI certification mark or performance classification.

8. Choosing appropriate lubricants for your car

In this chapter I make the presumption that you have read at least some of the earlier ones. You may find the terminology, and even some of the concepts, difficult to follow if this is not the case.

8.1 Engine
To choose the appropriate oil for your classic car's engine from the range of products available today, you have to make two major decisions – the most appropriate performance level for the oil and the most appropriate viscosity. Of course, this still leaves you with the choice of brand.

8.1.1 Choice of performance level
This choice depends, in part, on the original specification issued by the manufacturer of your engine, but it depends far more on the engine's recent history of use and servicing, and on what type of use you intend to put it to in the immediate future.

8.1.1.1 Prior use and servicing
Much has been written in magazines and online forums about the fact that modern oils contain detergents and older oils did not. It may be viewed as sacrilege but, in my opinion, this probably need not be a major concern to you.

How can this be when stories abound of detergent oils being put into cars that were not designed for them, and hellish problems resulting? Such an outcome is certainly possible if your engine has operated all of its life on low detergent oils. A modern oil will tend to clean surfaces and remove accumulated debris from oil seals. The debris may block an oilway, or it may reveal the fact that the effectiveness of the seals was, in fact, largely dependent on the accumulated debris.

However, unlike the situation 30 or so years ago – when many of these stories date from – there are very few cars today that are in the position of having spent their entire lives on 'vintage' oil. Non-detergent oils – oils of engine performance level SA or SB – have not been marketed generally since the early 1960s, although they have certainly been available from specialist suppliers. If your car has spent its operational life using such an oils, then it should continue to do so, at least until its next complete engine rebuild. At that time, unless the car is of such a vintage that it burns large amounts of oil in service, you could safely change to a modern oil, and to the benefit of your engine; it will wear less.

If your engine does burn large quantities of oil – and by large I mean of the order of a pint (500ml) or more in 60 miles (100 kilometres) – then an oil that contains detergents may exacerbate deposits in the combustion chamber(s). This is because the commonly-used detergents contain some inorganic chemicals and these do not

Which Oil?

burn; they form an ash. At normal rates of oil combustion the detergents in the oil look after the small amount of ash formed; they clean it up. However, at high rates of oil burning, the quantity of ash may be greater than the detergent can cope with. It is important to understand that I am talking here about engines that **burn** large quantities of oil. Many older engines consume large quantities of oil, but they do this mainly because of leaks.

If your engine has this inherent high oil burning characteristic – and it exists primarily with pre-1930 designs – then you should lubricate it with a low ash oil. This will be either a non-detergent oil or one formulated with ashless detergents. Check with the manufacturer of the oil, and also ask what its sulphated ash level is. This is the standard industry test or measure for ash.

8.1.1.2 Proposed use

It's an unfortunate reality that many classic cars are used very sparingly. Also, because it is known that complete lack of use is usually detrimental to machinery, many owners start their classic on a semi-regular basis, and run it until warm, either stationary (at idle) or by driving it a short distance. They then shut it down.

Although the practice is understandable and well-intentioned, this type of operation is just about the worst that can be imposed on an engine from the point of view of its lubrication. All of the engine starts are from cold, and after an extended shutdown period. The last semblance of oil has drained from all surfaces that are not submerged in the oil in the sump. The oil is cold and so it is viscous and slow moving. The engine almost certainly needs full choke (and often pumping of the accelerator) to get it to catch. Most of the operating period is in a fuel rich mode and, because the engine does not spend any significant time at high temperature, the excess fuel – fuel that has stripped lubricant from the bores as it drained down – is not boiled off from the sump. The oil is diluted. The same situation applies to the water formed by the combustion of the fuel. Some of it finds its way into the sump

---

Choice of lubricant brand

I am not going to recommend any one brand of lubricant over any other. However, I will strongly counsel against buying on the basis of lowest price unless this is absolutely essential for you.

It's an unfortunate reality that, although the various oil lubricant performance specifications are well-defined, and, at least for the past 20 years or so, also well-controlled, it is a fact that there are very few controls on the claims that marketers can make for their products. The situation is particularly difficult when the claims involve performance specifications that are now obsolete, and for which the required test procedures can no longer be run.

The best protection under these circumstances is to buy a product with a reputable brand. The rationale is that the marketer involved will recognise that its brand is valuable and will not indulge in practices that could damage it.

You should also beware of exaggerated claims. They may contain some grain of truth but they are unlikely to be worth the payment of a significant premium.

Choosing appropriate lubricants for your car

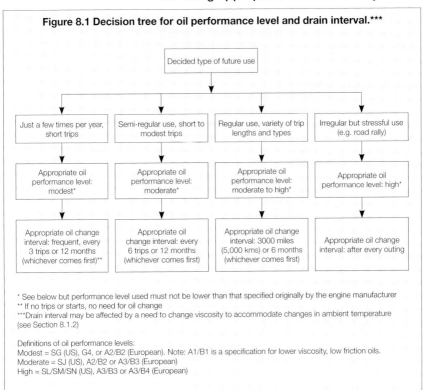

Figure 8.1 Decision tree for oil performance level and drain interval.***

* See below but performance level used must not be lower than that specified originally by the engine manufacturer
** If no trips or starts, no need for oil change
***Drain interval may be affected by a need to change viscosity to accommodate changes in ambient temperature (see Section 8.1.2)

Definitions of oil performance levels:
Modest = SG (US), G4, or A2/B2 (European). Note: A1/B1 is a specification for lower viscosity, low friction oils.
Moderate = SJ (US), A2/B2 or A3/B3 (European)
High = SL/SM/SN (US), A3/B3 or A3/B4 (European)

and so, when the engine is shut down, there is a load of water vapour waiting to attack the exposed ferrous parts.

For the sake of your engine, it would be much better to start it less frequently, but to drive it for much longer periods when you do. Even better, start it frequently and also drive it for long periods. Both you and your engine will benefit!

If you are this type of 'typical' user (short and non-frequent trips) then you can safely use an oil that has a modest performance level, provided it is from a reputable manufacturer/marketer. However, you should change it and the oil filter (if it has one) regularly. By regularly, I mean at least every three 'starts' or, subject to suitability of the viscosity (see Section 8.1.2 following), every twelve months, whichever comes first.

By modest performance I mean an oil of SG or similar performance (G4, or A2/B2 in European terminology). However, the performance level of the oil chosen must be at least equal to the one that the engine manufacturer recommended when the car was new. Using a high-performance level engine oil will be of no real benefit, and will simply cost you unnecessary money. Definitions of these general terms – modest, medium, high – are summarised at the bottom of Figure 8.1).

If you are an owner who does manage to do longer trips – say 30 minutes

Which Oil?

duration or more – but still only relatively infrequently, then again you can safely choose a moderate performance oil from a reputable source, and change it and the filter every six or so trips or every twelve months (whichever comes first, and again subject to variations in your local ambient temperature – see choice of viscosity following in Section 8.1.2). Again, the performance level must be at least equal to the one that the engine manufacturer recommended originally.

For both of these types of operation, regular oil changes are more important than the performance level of the oil used, provided, of course, that you do not use an oil with a performance level less than that specified by the engine manufacturer. There's no point in using synthetic oils (unless the climate dictates this – again, see the section on selection of viscosity that follows). The performance strength of synthetic oils is at high temperatures. They can, in fact, be inferior to mineral oils in the short trip mode of engine operation because they may not handle contaminants as well.

If you use your car regularly and for a variety of trip lengths and types – for example, if your car is your daily driver – then you should choose a higher performance oil, again from a reputable source. Do not buy on price as the main criterion. Subject to the provisos summarised in the panel on anti-wear performance, you can in general use oils of recent performance specifications, provided that the viscosity is appropriate (see following). Synthetic or part synthetic oils are acceptable. The oil/filter change interval should be about every 3000-5000 miles (5000-8000km), or six months, whichever comes first.

Finally, if the main use of your car is for semi-sporting use – that is, a limited number of periods of use but with high and extended engine stress levels during that use – then you should use a high-performance oil, and change it and the filter after every outing. Synthetic oils are a good choice for this type of use.

Anti-wear performance of the latest engine oils

To reduce exhaust catalyst poisoning, the maximum amount of phosphorus allowed in an engine oil has been restricted in the US and most of the western world since at least 1989. It was set at 0.12 per cent maximum with the introduction of the API SG engine oil classification. There was a further reduction to 0.10 per cent in 1996 with the introduction of SJ, and another recent reduction to 0.08 per cent in 2004 with SM.

In Europe, since 2004, this problem has been handled by introducing a special sequence of engine oils – the 'C' sequence oils (C1, C2, etc). These oils have low phosphorus levels and are also low in some other chemicals, such as sulphur. They are designed to be compatible with a variety of exhaust gas emission treatment systems.

As explained in Sections 4.3 and 4.5, phosphorus is added to engine oils as part of a chemical called zinc dithiophosphate (ZDTP), and this chemical is both an anti-wear additive and an antioxidant. There is currently some debate over the anti-wear performance of the latest, low phosphorus (API SM, SN Classification/ACEA C1, etc) oils in older cars. The topic is covered far more fully in Chapter 9 (Section 9.1.3). I suggest that you read that section before deciding which current performance level oil to use.

The decision is easier in Europe. The A sequence oils will do an excellent job in older cars, and so the C sequence oils can be easily avoided.

Choosing appropriate lubricants for your car

The steps in the suggested decision-making process to arrive at the appropriate performance level and drain interval for your engine oil are summarised in Figure 8.1.

8.1.2 Choice of viscosity

This is probably more critical than the choice of the performance level of the oil. You must choose the engine oil viscosity that is appropriate for your engine design, your local climate, and your personal pattern of use.

If your car dates from about 1970 or later – when multigrade engine oils were well-established – then the viscosity selection procedure is fairly straightforward. The manufacturer of your engine will have given all of the information that you need. However, for those whose vehicles are from an earlier era, when manufacturers couched their recommendations solely in terms of monograde oils – or even earlier, where they simply used terms such as 'thick' or 'thin' – things are not so simple.

To help such owners with this selection I have provided two methods – one for the monograde crowd (following) and one for the earlier 'thick/thin' group (see Section 8.1.2.1). The procedure to follow if starting from a monograde recommendation is summarised in Figure 8.2. There is also a set of charts to be used in conjunction with it (see Figures 8.19(xx) at the end of this chapter). There is a separate Figure 8.19(xx) for each of the three common SAE viscosity grades, SAE 30, 40 and 50. For example, Figure 8.19(40) relates to monograde SAE 40 engine oils.

To use the procedure, you will need the monograde viscosity recommendation

---

Synthetic oils, viscosity grades, and shear stability

It is most unfortunate that there isn't a clear, mandatory definition of what constitutes a synthetic oil. First, there's the problem outlined in the panel in Section 3.1, whereby base oils that are made by entirely chemical means, and base oils that are made by extensive chemical and physical refining of natural mineral oil can both be called synthetic. Second, is the fact that there is no policed definition of what can be called a synthetic oil in the marketplace. An oil blended with only a percentage of synthetic base stock – the rest being conventional mineral oil – may be described with impunity by its marketer as synthetic.

Why is this important? At one level it is not. Provided that the product meets the performance specifications and viscosity grades claimed all will be well. However, as you read this chapter, you will learn that the shear stability of an engine or gear oil – that is, its ability to retain its initial viscosity while in use – is critically important. There is a natural tendency to presume that a synthetic oil will be completely shear stable. This is because many synthetic base oils have inherently high viscosity indexes and there is no need to add polymeric VI improvers to them. It's the polymeric substances that are shear unstable. However, if a multigrade oil product is actually a mix of synthetic and conventional base oils, it will almost certainly also contain a quantity of VI improver in order to meet its viscosity grade. It will not necessarily be shear stable.

So, unfortunately, to ensure that the oil you are considering has the shear stability level that you require, you'll have to obtain this information (the shear stability index, see Section 4.1.1) from its marketer or manufacturer. You cannot simply presume that a product described as synthetic will be shear stable.

Which Oil?

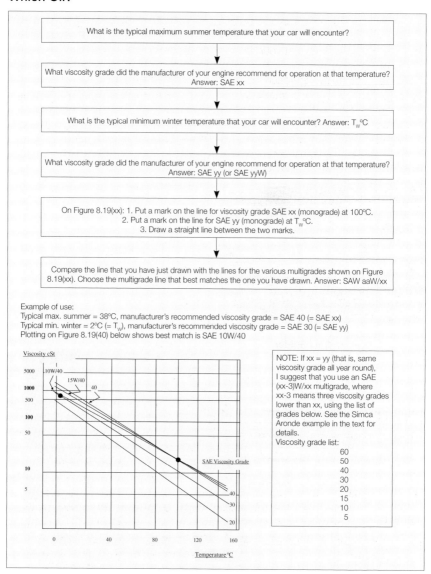

Figure 8.2 Decision tree for selection of engine oil viscosity if manufacturer recommended monogrades.

given by the manufacturer of your engine when it was new, and the typical annual minimum and maximum ambient temperatures in the area you plan to use your

Choosing appropriate lubricants for your car

car. By typical, I do not mean the average (this gives the average of high and low days). What you need is the minimum temperature in Winter and the maximum temperature in Summer that your car could see reasonably regularly if it were in operation.

To demonstrate the approach that I recommend for selecting the appropriate viscosity, I will give a few examples based on my own situation and that of a friend. I own three older cars – a 1979 Lancia Beta Spider (also called a Zagato), a 1969 Porsche 912 (also called the poor man's 911 – part of the label being true in my case) and a 1936 Austin Seven (the people's car). My friend owns a 1925 Rolls-Royce Twenty (the car of the other sort of people). Also, to partially fill the gap between 1936 and 1969, I will include a 1961 Simca Aronde that I once owned briefly.

Lancia Beta Spider.

The vehicle that I use almost daily is the Lancia. It has a four-cylinder, twin overhead cam engine that is still quite modern in its design. All of the engine lubricating oil is filtered via a screw-on, disposable, paper element filter of the type that is still generally current today.

The Lancia is from an era when multigrade engine oils were well-established, and so I do not have to make use of the procedure in Figure 8.2. Its manufacturer (Fiat/Lancia) specified the appropriate engine oil viscosity to use in a very helpful and comprehensive way, as reproduced in Figure 8.3.

I live in Sydney, Australia, and the typical minimum temperature in winter is 2°C (36°F) while the typical summer maximum is 38°C (100°F). You will see that, for a maximum temperature of 38°C Fiat/Lancia recommends either an SAE xW/50 multigrade or a straight SAE 40

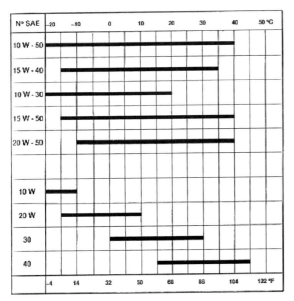

Figure 8.3 Fiat/Lancia viscosity recommendations for 1979 Lancia Beta.

87

Which Oil?

monograde. You might wonder why it didn't specify SAE xW/40 multigrades as well. Presumably it was aware that the multigrades available in 1979 were not necessarily very shear stable. They could easily shear down one viscosity grade.

We can conclude that the viscosity that Fiat wanted in an environment at 38°C ambient was an SAE 40. This could be achieved either via a monograde, or by a multigrade that started life one grade higher. Thus, from consideration of the summer temperature, I have reduced the possible choice of engine oil viscosity to four grades – 10W/50, 15W/50, 20W/50 or straight 40.

If I look at the other end of the spectrum – my 2°C winter minimum – and now restrict my consideration to only the four oils that can handle the summer maximum, I am immediately reduced to three oils; the SAE 40 monograde is ruled out. Its minimum recommended operating temperature is 15°C, well above my 2°C. I am left with a choice between three multigrades – 10W/50, 15W/50 and 20W/50. In my circumstance, the 20W/50 gives plenty of margin at the low temperature end (it is recommended down to minus 1°C). Given that this viscosity grade is readily available in the area where I live, that it comes in a variety of performance levels, and that, when formulated from conventional mineral base stocks, it is far cheaper than the other two alternatives, this is the viscosity grade that I choose to use.

In Figure 8.4 I have plotted the viscosity of an SAE 20W/50 oil and compared it with both a monograde SAE 20 and a monograde SAE 50 oil. In looking at such plots you need to consider the scales on the axes carefully. It may seem that the multigrade SAE 20W/50 is little different to the monograde SAE 50. However, at my minimum temperature of 2°C, its viscosity is about 1500 cSt, or 33 per cent lower. It has the same viscosity at 100°C, and is about 1 cSt or 15 per cent higher in viscosity at 140°C, where my engine will benefit from the greater protection.

From Figure 8.4 you will note that the SAE 20W/50 viscosity oil has a viscosity that is far higher than an SAE 20 oil, even down toward 0°C. However, we have to be careful at low temperatures. We cannot simply continue to extrapolate the graphs. Wax will start to play a part, and change the viscosity/temperature relation. What we do know (from the SAE viscosity grade definitions, Table 5.6), is that both of these oils will have the same viscosity in the range -15 to -20°C.

Fiat was very

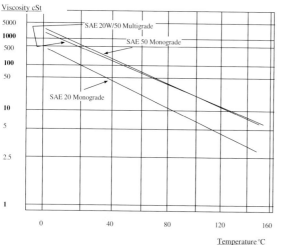

Figure 8.4

Choosing appropriate lubricants for your car

comprehensive with the information that it gave for my Lancia, and also, by 1979, multigrade oils were well established. The situation was similar in this respect to that which prevails today, and so the task of translation was quite simple. Things tend to become less straightforward as we go back further in time.

My second car, the 1969 Porsche 912, was built 10 years earlier than the Lancia.

Expected ambient temperature	Appropriate viscosity
Below +5°F (-15°C)	SAE 10
From +5 to 32°F (-15 to 0°C)	SAE 20
From 32 to 86°F (0 to 30°C)	SAE 30
Above 86°F (30°C)	SAE 40

Figure 8.5

Again, I use it quite regularly, at least once a week, and I drive it all year round. The engine is derived from the air-cooled VW design and, while there is an engine oil cooler, the filtration system is the by-pass type via a paper element filter that sits in a canister.

The Porsche recommendation from 1969 is given in Figure 8.5. Although multigrade oils were in widespread use by 1969, Porsche was not recommending them. It presumably had misgivings regarding shear stability, particularly given the fact that its cars were sports cars and could live quite stressful lives.

For my local temperature extremes of 2°C in winter and 38°C in summer, and if I were to follow Porsche's lead in avoiding multigrades, I would have no choice but to change the viscosity grade of my engine oil twice each year. To comply with the Porsche recommendation I would have to use an SAE 40 in summer and an SAE 30 in winter. I would also have a concern that the SAE 30 might be borderline in mid-winter, as it only just handles my 2°C minimum.

Happily, today, I can simplify my life and also do a better job of lubricating the 912's engine. This is because I have access to multigrade oils of good shear stability. To help me with this decision I will use the procedure in Figure 8.2.

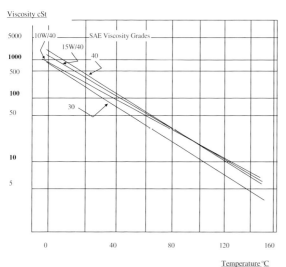

In my situation, using the information I have given above, and following the flow chart in Figure 8.2, I arrive at TW=2°C, xx=40 and yy=30. I make the appropriate marks on Figure 8.19(40) and get the outcome that is shown as the example at the bottom of Figure 8.2. Following the procedure results in a suggestion to use an SAE 10W/40 oil.

In Figure 8.6 I have reproduced Figure

Figure 8.6

Which Oil?

Porsche 912.

8.19(40). It shows the viscosities of the two relevant monograde oils – the SAE 30 and SAE 40 recommended by Porsche in 1969 – against temperature. I have also plotted a 10W/40 and a 15W/40 multigrade oil.

In looking at the plots we should consider the fact that the temperatures at the critical locations within the engine, once fully warmed up, will be in the range 100 to 140°C. The sump temperature will be lower than this because of cooling, and because some of the oil is lubricating lower temperature areas, such as cams. Also, in the case of the 912, there is an oil cooler. In considering the 'hot' viscosity, it's the temperature of the oil in the locations where it has to do the lubrication that is the critical issue, not its temperature in the sump.

When we consider this, a second advantage of the multigrades (in addition to their lower viscosities at low temperatures) becomes obvious. In the critical 100-140°C operating range, they have a small but significantly higher viscosity than the monograde SAE 40 and a much higher viscosity than the SAE 30 grade.

Considering winter operation, Porsche specified the use of an SAE 30 viscosity grade to handle cold starts during my winter temperature of 2°C (also see the panel on the interpretation of winter viscosity grades). You will see from the plot that the SAE 10W/40 multigrade has a viscosity very similar to an SAE 30 oil at 2°C. Given that it is also provides at least equal or better engine lubrication at warmed-up operating temperatures as the SAE 40 oil that Porsche specifies for my summer use, it is apparently the perfect choice for year round motoring under my ambient conditions. My only potential concern is shear stability.

Today, all 10W/40 oils marketed will be fully or partially synthetic. Use of a semi or wholly synthetic oil in the 912's engine would not be a concern to me (other than the cost!). It has the advantage that the formulator of the oil will have used very little (if any) polymeric viscosity index improver, so shear will not be an issue. Less polymer is also of potential benefit from a deposit perspective in a hotter, air-cooled engine. However, as I use the car only once or twice a week, and usually only for fairly short trips, I

Interpretation of engine oil viscosity recommendations at low temperatures

Car manufacturers are forced to specify lower viscosity engine oils in colder ambient conditions simply to handle the cold start and initial cold operation periods. They would really prefer you to use a higher viscosity.

How can I make this statement? The reality is that, once an engine is fully warmed up, the ambient air temperature has only a relatively minor effect on the temperatures deep within it, where the most critical lubrication tasks are happening and where full-film lubrication – which is dependent on the oil viscosity – is needed.

Fortunately, today we can get closer to the ideal by using multigrade oils.

Choosing appropriate lubricants for your car

choose to use a conventional 15W/40. I feel that it will handle my particular operating environment a little better than a synthetic, and I err on the side of frequent oil changes. If my use pattern changed to include longer trips I would move to a 10W/40.

You may feel that I have been a little cavalier in deciding to use the 15W/40 viscosity grade instead of the 10W/40. I may be taking a risk with respect to its winter performance. In reality, I am not. The method given in Figure 8.2 gives a conservative (that is, very safe) outcome for the 'W' rating. You can see this by comparing Figure 8.6 and Figure 8.4. Fiat/Lancia was happy to recommend a 20W/50 for my circumstance, even though its viscosity at 2°C is well above that of a 20 grade oil at this temperature.

We now journey from 1969 back to 1961 and the Simca Aronde. It has a four-cylinder engine that was quite advanced for its time. The engine lubrication system is fully pressurised, with all of the oil passing through a centrifugal filter before flowing to the bearings, etc. This filter did not contain a filter element, and required cleaning only every 35,000 miles (55,000km).

Simca Aronde.

Although the Simca was a French car, in Australia, where I live, it was manufactured, sold and serviced by Chrysler Australia. Chrysler required the use of an engine oil that had both an SAE viscosity designation and an MS (Medium Service) engine performance classification. 1961 was prior to the formalisation of the API engine oil performance classifications – the 'S' series. MS was equivalent to what became known as SC (see Section 5.1.1).

Chrysler/Simca was comprehensive in its specification of the engine oil viscosity, as shown in Figure 8.7. Multigrade viscosity oils were relatively new in 1961, and Chrysler gave the owner the option of monograde or multigrade. The multigrades at the time were quite narrow grades, by today's standards, and they were not necessarily highly shear stable.

For the ambient temperature range in Sydney (2 to 38°C), the Chrysler recommendation allows the use of an SAE 30 oil year round. Applying the procedure in Figure 8.2 (including the Note), the suggested oil viscosity to use is SAE 10W/30. This suggestion corresponds to one of Chrysler/Simca's own multigrade

Anticipated temperature range	Recommended viscosity grade	Optional multi-viscosity grades
Above 32°F (0°C)	SAE 30	SAE 20W/40 SAE 10W/30
As low as +10°F (-12°C)	SAE 20W	SAE 10W/30
As low as -10°F (-23°C)	SAE 10W	SAE 10W/20

Figure 8.7 Chrysler/Simca 1961 viscosity recommendations.

Which Oil?

recommendations. However, if I still had my old car, with its fairly well used engine, I would probably move up a grade and use an SAE 15W/40.

My remaining car takes us further back in time. It is a 1936 Austin Seven which I use quite infrequently and then only on short trips. (It is actually a car that I built up from parts many years ago – when I was a student. The body is the remains of a 1929 Meteor Sports, but the powertrain is from a 1936 Ruby.) Austin's viscosity recommendations from 1936 are summarised in Figure 8.8. I have to admire its frankness and the confidence that it had in its product. Would it really be wise to go driving in the Arctic in an Austin Seven with a worn engine?

The recommendations are very straightforward. For me in Sydney – not quite in the tropics but a little warmer on average than Britain – I should probably run SAE 40 in summer and SAE 30 in winter to satisfy Austin. However, again, a multigrade can save me the hassle and do a better job of looking after my engine.

It happens that, for me, the Austin recommendation is exactly the same as the Porsche one, and so the outcome from using the procedure in Figure 8.2 is a suggestion to use an SAE 10W/40 oil. Again, I actually choose to use a 15W/40. If I lived in Britain, where Austin specified an SAE 30 for year-round use, then, following the logic given in the note in Figure 8.2, I would use a 10W/30 oil. The situation and outcome are the same as applied with my Simca Aronde.

Both of these grades – SAE 15W/40 and SAE 10W/30 – can be obtained as conventional mineral oils, and these would be a perfectly satisfactory, economical choice for this application (the 10W/30 in the UK and the 15W/40 in Australia). An Austin Seven engine is not a particularly demanding environment.

Finally, we go back another 11 years to the 1925 20HP Rolls-Royce that has been owned for many years by a friend.

Austin 7 Meteor.

Engine condition	Area of operation	Viscosity grade
Good	Tropics	SAE 40
Good	British Isles (summer and winter)	SAE 30
Good	Arctic	SAE 20 or 10
Worn or sports engine	Tropics	SAE 50
Worn or sports engine	British Isles (summer and winter)	SAE 40
Worn or sports engine	Arctic	SAE 30 or 20

Figure 8.8 Austin's 1936 viscosity recommendations.

Choosing appropriate lubricants for your car

It also resides in Sydney and so it faces the same ambient temperatures, a range from about 2 to 38°C.

The actual recommendation given by Rolls-Royce is a little difficult to deduce. My friend has several apparently-genuine, period handbooks. In the one that was issued in November 1925 – the year that the car was built – the recommendations for the engine oil are:

Price's Motorine 'B' for summer
Price's Motorine 'C' for winter

The handbook also advises that Rolls-Royce Ltd can supply any quantity of these oils at current retail prices, with free delivery in the London area. For other locations, orders of five gallons or more will be delivered free to the nearest railway station.

Before testing Rolls-Royce at its word regarding delivery of Price's Motorine, my friend has a second, also-apparently-authentic handbook. It states that it incorporates the instructions given in both the November 1925 and April 1927 editions, and also Edition No 7 (unfortunately not dated). Presumably, it was issued sometime late in the 1920s. The recommendations that it gives for the engine oil are shown in Figure 8.9. More brands have been introduced but, somewhat strangely, the Price's Motorine grades have been changed – the 'C' viscosity is now recommended for summer, not winter, and a different grade – 'M' – is the new winter recommendation.

Rolls-Royce Twenty.

Brand/product	Summer	Winter
Price's Motorine	'C'	'M'
Wakefield's Patent Castrol	'XXL'	'XL'
Vacuum Mobiloil	'BB'	'A'
Shell	'Triple'	'Double'
Duckham's Adcoidised	'NP3'	'NPXX'

Figure 8.9 R-R's engine oil recommendations c1929.

Brand/product	Summer	Winter	Worn engine
BP Energol	30	30	40
Wakefield's Castrol	'XL'	'XL'	'XXL'
Shell	X-100 SAE 30	X-100 SAE 30	X-100 SAE 40

Figure 8.10 R-R's engine oil recommendations c1931.

However, my friend's archives are not yet exhausted. He has a third authentic handbook, now combining April 1924 (a new addition to the list), November 1925, April 1927, Edition No 7 (still undated) and Edition No 8 (also undated). Its recommendations are shown in Figure 8.10.

We are now looking at a handbook issued

Which Oil?

by Rolls-Royce presumably in the early 1930s, and here, for the first time, we see reference to the SAE viscosity grading system. This system, in the form quoted by R-R, was introduced in the US in 1926 (see the panel in Section 3.1). Hence, it had crossed the Atlantic by 1930 or so.

By this date, Rolls-Royce had also revised its engine oil recommendations for my friend's 20HP. In 1925, when it was built, R-R specified a lighter oil for summer than winter. However, by 1930 or so, it was recommending that an SAE 30 viscosity grade be used all year round. Presumably it had more confidence in the viscosity characteristics of the oils that had been developed in the five or so years since the car was launched and that complied with the new SAE viscosity grade system.

I have devoted quite a lot of time to the recommended engine oil for this Rolls, but I feel that it's both interesting and educational. You can see the rapid technical progress that was being made in lubricating oils in the late 1920s, and also the introduction of standardised viscosity specifications. You can also see Rolls-Royce's willingness to move with the times and to modify its recommendations as the improved lubricants were developed.

Given that the R-R recommendation of SAE 30 engine oil all year round is the same as Chrysler/Simca's 1962 recommendation for the Aronde, the outcome from using Figure 8.2 is the same. The suggestion is to use an SAE 10W/30 viscosity grade. Again, if it were my car, I would probably use an SAE 15W/40. Summer in Sydney may be hotter than the summers that R-R had in mind in 1931, with or without allowance for global warming.

8.1.2.1 Vintage and veteran engines

Many vintage and veteran cars were produced before standard systems for definition of oil viscosity were developed and adopted. Manufacturers typically used imprecise descriptive terms such as 'thick' or 'thin.' The owners of such vehicles potentially cannot follow the method outlined in Figure 8.2, because they cannot answer the question "what viscosity grade did the manufacturer of your engine recommend for operation at that temperature?". Standardised viscosity grades were introduced in 1911, initially in the United States (see Section 3.1). However, it took 15 to 20 years for them to become universally accepted.

If you do not know the viscosity grade recommended initially for your engine, the simplest (and probably the safest) approach is to base your choice on what other owners of identical vehicles have found to be successful. However, if no such advice is available, the following method can be used as a guide. To use it you will need to know some mechanical details of your engine.

In engine lubrication – presuming the design does not have some specific individual idiosyncrasy, flaw or weakness – the critical requirement is maintaining full film lubrication of the crankshaft and big end bearings (see Section 3.5), presuming these bearings are the plain type. You may imagine that vintage and veteran engines were relatively understressed. However, this is not necessarily the case. On some engines, bearing sizes were relatively small for the piston sizes and loads generated.

The load on the engine bearings comes from three main sources: the downward

Choosing appropriate lubricants for your car

force on the piston arising from the combustion of the fuel, the inertia force at the top and bottom of the stroke as the piston changes direction of travel, and the rotating inertia as the crankshaft and attached big ends rotate. These forces vary during the engine rotation, sometimes reinforcing each other, sometimes opposing. The typical consequences during the cycles of a four-stroke engine are shown in Figure 8.11. Valve timing is also important in determining the details of this curve.

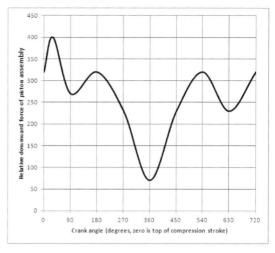

Figure 8.11 Variation of load on bearing (typical).

The greatest bearing load occurs in the early part of the expansion stroke, immediately after the fuel charge is ignited, and its magnitude is determined by a combination of the compression ratio of your engine, the amount of fuel that it burns to generate the expansion stroke, and its rotating speed. Valve timing, cam profile, inlet tract design, carburettion, etc, all influence the actual compression achieved by an engine, but they are secondary, albeit important, effects when compared with the mechanical compression ratio.

The mechanical details of your engine that you will need to know if you wish to apply the method that follows are:
V – its volumetric capacity (litres or cubic inches)
P – its maximum power output (KW or HP) at
N – revs (revolutions per minute),
C – its compression ratio
B – its cylinder bore diameter (mm or inches)
L – the length (sometimes called the width) of its big end bearings (mm or inches)
d – the internal diameter of the big end bearings (mm or inches, also equal to the diameter of the crankshaft pins)

Having gathered this data, calculate the two load values below. Make sure that you use consistent units – either **all** metric or **all** imperial/American – and that you use the appropriate formula for LV2 for the units you have chosen. Your answers should be numbers in the range of about 1 to 7. If your values fall outside this range then you have probably made a mistake. Check again.

Load Value 1 (LV1) = $(C/9)^{1.4} \times B^2 / (L \times d)$
Load Value 2 (LV2) = $100 \times P \times B^2 / (V \times N \times L \times d)$ (Metric)
Load Value 2 (LV2) = $4550 \times P \times B^2 / (V \times N \times L \times d)$ (Imperial)

Which Oil?

Load Value 1 is related mainly to the load generated by compression. Load Value 2 brings into account combustion and the more specific features of your engine, such as its efficiency.

For many British cars the engine capacity is quoted in litres, but all other units are Imperial. Convert the litres to cubic inches by multiplying by 61. Also, to help you to calculate $(C/9)^{1.4}$, simply read its value from Figure 8.12 for your particular compression ratio C (you may have to interpolate).

Armed with the two Load Values for your engine, use Figure 8.13 to find the oil viscosity that will give full film lubrication of your big end bearings.

Note that the viscosity value obtained from Figure 8.13 is for moderate ambient conditions (typically 20 – 25°C). You should make allowance for your own climate. In the spirit of the times when your engine was manufactured, use a thicker oil if it is hotter and a thinner one if colder. Also, strongly consider using a multigrade oil – SAE10W/30 if Figure 8.13 indicates SAE30, 15W/40 for SAE40, and 20W/50 for SAE50.

For older, relatively low RPM engines with large cylinder bores, you may find that the load values calculated are high (greater than 4), and that the oil viscosity recommended is SAE50 or even thicker (off the scale). This gives a practical problem. If such a thick oil were used, engine starting when cold would be very difficult, and viscous losses in operation from start up would be high.

The high bearing loads

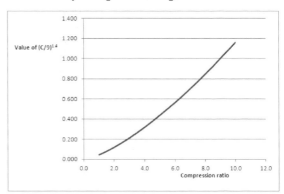

Figure 8.12 Compression ratio factor.

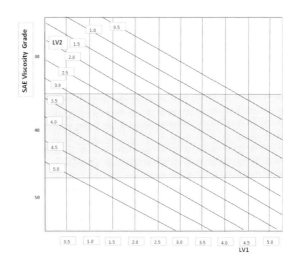

Figure 8.13 Estimated viscosity grade.

Choosing appropriate lubricants for your car

in these engines are real. The best approach would be to use a wide multigrade oil, at least a 20W/50 or wider.

Finally, the method applies only to older engines. As machining tolerances and surface finishes have improved over time, it has become possible and desirable to use less viscous oils.

8.1.2.2 High viscosity engine oils

Until recently, the highest viscosity grade sold generally for engine oils was SAE 50. Today you'll find oils that are SAE 60 and even higher (even though there is no current official definition of an engine oil viscosity grade higher than SAE 60, see Table 5.6).

You should be cautious about using these high viscosity oils, particularly if they are monogrades or only narrow multigrades. They were developed in the high-performance, large engine arena – typically US-style V8s running in drag races or the street equivalent. These engines are not high revving. They have large loads but they are acting on large bearing areas. Also, the engines often have relatively large bearing clearances to ensure high oil flows. Long service life is not an issue for these engines, surviving the short term bursts of extreme power and torque is. In this circumstance a low viscosity oil will tend to be displaced more easily from the bearing areas and a higher viscosity will help.

However, if your engine and usage pattern are not of this type, a high viscosity oil will not be beneficial and may very well be detrimental. For example, your engine may require a high oil flow through critical areas to take away heat. The more viscous the oil, the lower its flow rate.

8.2 Transmission

Here, I'll be covering automatic transmissions separately (Section 8.2.3). For all other transmission components (gearboxes and final drives) the selection task is essentially the same as it is for your engine. To choose the appropriate oil (or oils) from the range of products available you have to decide two things – the most appropriate performance level for the oil, and the most appropriate viscosity.

8.2.1. Choice of performance level

All gear sets in all transmissions, irrespective of their vintage, will spend some time in the non-full-film lubrication zone during operation. They therefore require (or will benefit from), some level of anti-wear/extreme pressure capability in their lubricants. The critical question is what level? Referring to Table 6.2, is it GL-3 or is it GL-4/GL-5?

If you have no other information, the best assumption that you can make is that the gearbox – if it is a separate unit with its own lubricant – will require at least GL-3. The final drive – if it is a separate unit with its own lubricant – will require at least GL-4 and probably GL-5. This latter assumption (GL-5), can be upgraded to certainty if the final drive is a hypoid design.

If your vehicle has a transaxle in which the gears and the final drive share the same lubricant (which is the case in most transaxle designs), things are less

Which Oil?

> **EP caution with some British cars**
>
> I have stated in the main text that, if a manufacturer specifies Extreme Pressure gear oil for its gearbox or differential, then you are safe to use such a product. The manufacturer will have ensured that there are no metals in the box that could be attacked by the EP additives.
>
> Unfortunately, this logic cannot be applied universally. In the years immediately following the Second World War, many major British manufacturers specified EP oils for their gearboxes, and yet they contained metals such as bronze.
>
> I can only assume that the manufacturers concerned simply did not fully understand the chemistry of EP oils. As was explained in Section 6.1.2, these oils were developed during the war mainly under the auspices of the US military. They were not taken under the wing of the oil industry until the early 1950s. The British car manufacturers were perhaps not fully in the loop.
>
> If you have a British car manufactured between about 1935 and 1955, and it specifies EP oil for the gearbox, be cautious. Check what other owners of your car are using (the problem is reasonably well understood in enthusiasts' circles). If no such advice is available, check whether the gearbox has bronze bushes or synchromesh rings. If an EP oil must be used then choose one whose manufacturer states that it is safe with vulnerable metals.

straightforward. While most manufacturers require GL-3 performance, some require GL-4/GL-5. If your vehicle has a transaxle, do your utmost to find out the original lubricant specification for it. Fortunately, in mass production cars, transaxles are a relatively recent development, so this information is generally fairly readily available, at least as far as which lubricants (by name) were recommended at the time.

Presuming that at least one of the oil companies that was originally named is still trading, I suggest that you contact their current technical department with the name/model/year of your car and the name of their oil that was recommended for it historically. They should be able to give their current recommendation. If they do this by quoting a product, insist on them also giving you its technical data sheet. This will give the performance level (as well as other useful information) and allow you to consider other brands of lubricants.

To assist you to follow the logic outlined above, Figure 8.14 gives a summary. Also, to give you some examples of selecting the appropriate performance level for transmission lubricants I will again use my own cars, plus one owned by a friend. To remind you, they are a 1979 Lancia Beta Spider, a 1969 Porsche 912, a 1961 Simca Aronde, a 1936 Austin Seven, and a 1925 Rolls-Royce Twenty.

Starting with the Lancia, it is front-wheel drive, has a manual gearbox, and a transaxle final drive. The gearbox and transaxle share the same lubricant.

In the 1979 driver's handbook, Fiat/Lancia specified the use of any one of the following oils:
Agip F1 Rotra SAE 85W/90
Esso Gear Oil GX 85W/90
Mobil Lubrite LZ90
OlioFiat ZC 90 SAE 80W/90

I also have an aftermarket workshop manual for the car. It was published in

Choosing appropriate lubricants for your car

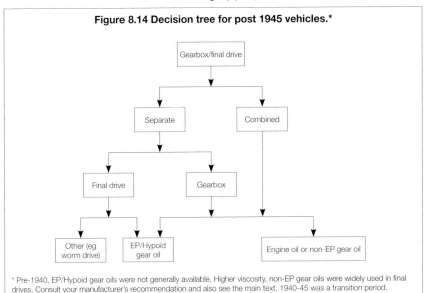

Figure 8.14 Decision tree for post 1945 vehicles.*

* Pre-1940, EP/Hypoid gear oils were not generally available. Higher viscosity, non-EP gear oils were widely used in final drives. Consult your manufacturer's recommendation and also see the main text. 1940-45 was a transition period.

1981 and specifies the use of Castrol GP50 (presumably, there was a commercial arrangement between the publisher of the manual and Castrol and so, unfortunately, no other oil is listed).

Although all of these oils were available in the early 1980s this is not the case today. Nor are their performance specifications available (at least not readily).

Worm drives

Some vehicles in the past have used a worm gear for the final drive, perhaps most notably the Peugeot 404. In modern vehicles, the Torsen differential uses a version of this drive.

Worm drives can give large gear ratios, are quiet, and can withstand heavy loading. They are still used in some forms of heavy vehicles, including some tractors.

A disadvantage is the friction caused by the sliding motion. To reduce it, it's common to use a phosphor-bronze worm wheel. If this is the case then conventional EP gear oils cannot be used because they will corrode the phosphor-bronze (see Section 6.1.2). Oils with oil-type anti-wear additives are often specified (Section 4.3). If your vehicle has a worm drive rear end then you should follow its manufacturer's recommendation closely when choosing a lubricant.

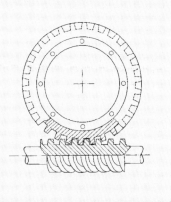

Which Oil?

None of the products has the letters EP in its name. Also, from the way that its viscosity is specified (SAE 50), the Castrol product was almost certainly an engine oil. I conclude – admittedly from somewhat circumstantial evidence – that Fiat/Lancia did not require the use of an extreme pressure (GL-4/GL-5 performance oil) in the gearbox/final drive; it required GL-3 performance. To be safe with my conclusion I contacted Castrol who confirmed that GP50 had been an engine oil (in fact, it's still marketed in some regions under Castrol's Classic Oils range). My final choice of lubricant will depend on viscosity considerations (see Section 8.2.2 following).

Turning now to my Porsche 912, it also has a combined manual gearbox/final drive, albeit of a very different design and layout to the Lancia. Back in 1969 – again in the driver's handbook – Porsche gave a very simple specification for the lubricant to be used: SAE 90 hypoid gear oil. As was explained in Chapter 5, hypoid gears require an oil with at least some EP performance for their lubrication. Thus, unlike Lancia, Porsche required the use of at least a GL-4, and probably GL-5, performance lubricant in its transmissions.

At the time (1969) that my Porsche was new, the types of EP additives used in hypoid gear oils to give GL-4 and GL-5 performance were sulphur-based, and chemically quite aggressive to copper, bronze, etc. The 912's gearbox has Porsche's own patented synchromesh set up on all forward gears. The fact that it specified an EP oil indicates that it was comfortable with the use of these aggressive chemicals in the lubricant. The boxes had appropriate metallurgy (no 'yellow' metals).

I'm fortunate that, when choosing a GL-4/GL-5 gear oil today, I have more choice than was the case in 1969. For my Porsche I use a product that claims to be suitable for most manual gearboxes as well as hypoid differentials. It will have less-aggressive chemistry than was available in 1969. However, it must (and does), have EP performance, not just anti-wear, and it's of GL-5 performance level.

We now move on to the 1961 Simca Aronde. Fortunately, Chrysler/Simca gave a very simple and direct recommendation – EP gear oil for both the gearbox and the rear axle differential. Thus the appropriate performance level is GL-5.

Moving back to 1936, for my Seven, Austin specified engine oil for its gearbox, and Extreme Pressure gear oil for the differential. Again, these are straightforward, unambiguous specifications, ones that I can easily comply with today.

Finally there is the 1925 20HP Rolls-Royce. I would like to spare you the full saga of the three handbooks if I could, but unfortunately the recommendations again vary issue to issue. In the original 1925 handbook it's Price's Amber 'B' Gear Oil for both the gearbox and the back axle. This is apparently a different product to the Price's Motorine 'B' recommended for the engine in summer. The common 'B' terminology probably indicates similar viscosities of the two products.

This is confirmed in the next (sequential) handbook which recommends the summer (not the winter) grade engine oil for the gearbox (as per Figure 8.9) and Price's Amber 'A' for the rear end, mixed with up to 10 per cent Price's Motorine 'C' in winter.

Choosing appropriate lubricants for your car

Turning to the 'last' handbook – dating from about 1930 – the recommendations now become:
BP Energol 90
Wakefield's Castrol ST
Shell X-100 SAE 60
for the gearbox, and the same products for the rear axle, but with viscosity 80/90.

What is constant throughout is that an engine oil is suitable for both applications – gearbox and final drive – albeit with the recommended viscosities changing somewhat over time.

No EP or anti-wear performance or additive is specified for either the gearbox or the rear axle. This is not surprising as the car predates their widespread development and use. In theory, straight mineral oil would be suitable.

If it were my car, I would use an engine oil with good anti-wear performance – one of GL-3 performance level – in both the gearbox and the rear axle. Viscosity selection is covered in the next section.

8.2.2 Choice of viscosity

For transmission oils, the viscosity is almost invariably expressed using the SAE system (Section 6.1.1). However, you need to be alert to the fact that there are two SAE viscosity scales, one for engine oils (SAE J300, Table 5.6), and one for gear oils (SAE J306, Table 6.1). Although the gear oil scale is obviously the one of most relevance here, it is a fact that, for manual gearboxes, some manufacturers have specified the use of engine oils in the past (my Austin Seven is a good example). If they have, then the viscosities they refer to will be engine oil viscosities. The relationship between the two scales is shown in Figure 8.15.

Multigrade viscosity gear oils are also available. Referring to Figures 8.15 and 8.16, products with viscosity grades such as SAE 75W/90, 80W/90 and 85W/140 are available. As was explained in Section 6.1.4, some of these grades require the use of either a viscosity index improver or a synthetic base oil to achieve the required viscosity/temperature characteristics. If a viscosity index improver is used in a gear oil it must be highly shear stable – an order of magnitude more stable than the VI improvers generally used in engine oils. While such highly shear stable VI Improvers are available – for example, ones based on the class of polymers called polymethacrylates – it's safe to assume that multigrade engine oils are not suitable for gearbox application if they are formulated with polymeric viscosity index improvers, rather than with synthetic base stocks with inherently high viscosity indexes.

The two most common viscosity grades marketed for transmission oils are SAE 80W/90 and 85W/140. Again referring to Section 6.1.4, the relative narrowness of the multigrade ranges in these two products means that, when using a modern, highly refined, mineral base oil, they can be formulated without the need for the addition of a viscosity index improver or the use of any synthetic base stock. However, to achieve this outcome, these products have to be formulated at the very lowest end of the quoted viscosity range (the 90 and the 140 ranges respectively).

Which Oil?

If the manufacturer of your transmission specified the use of either of these multi-viscosity grades – SAE 80W/90 or 85W/140 – then this 'low end' viscosity need not be a concern. The manufacturer would have been aware of the situation and, apparently, was comfortable with it. If your transmission predates the introduction of these multigrades (the early 1960s) then, as you'll soon see, you'll have to do a little more thinking to choose a lubricant.

I will again use my own cars to give you an idea of the approach that I suggest you use to choose the appropriate viscosity grade for your vehicle. Starting with the 1979 Lancia Beta Spider, as was listed in the previous section, Fiat/Lancia specified the use of any one of the following oils for its gearbox/transaxle:

Agip F1 Rotra SAE 85W/90
Esso Gear Oil GX 85W/90
Mobil Lubrite LZ90
OlioFiat ZC 90 SAE 80W/90

My aftermarket workshop manual for the car, published in 1981, specifies the use of Castrol GP50, an engine oil.

The first thing to note is the variation in the viscosities of the listed alternatives. There are apparently four different viscosity grades (in reality there's only one). If you look at Figure 8.15 (using the old scale for gear oils, the recommendations date from pre-2005), you'll see that the SAE 50 engine oil viscosity range falls right in the middle of the SAE 90 gear oil range.

The two SAE 85W/90 oils would also sit in the middle of the range. 1979 was still in the early days of 'multigrade' gear

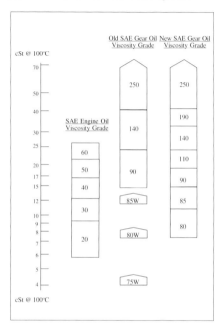

Figure 8.15

Multigrade gear oils and VI improvers

In the text, it is explained that, with modern mineral (non-synthetic) base oils, it is possible to formulate some multigrade gear oils without any need to add a viscosity index improver. Such products are, therefore, inherently shear stable. The multigrade ranges (under the pre-2005 viscosity classification system) that can be achieved without VII are:

SAE 80W/90, 85W/90 and 85W/140

The 'range' of the multigrading is relatively narrow and not of as much benefit as a 'genuine' multigrade. For these, addition of VI improver, or the use of at least some synthetic base oil, is needed.

Choosing appropriate lubricants for your car

oils. Apparently, Agip and Esso were not yet ready to lower the viscosity of their products so that they could claim SAE 80W/90. They kept them in the centre of the SAE 90 range and, as a result, could only claim SAE 85W as the cold temperature viscosity.

We can conclude that all five oils are SAE 90, with four of them having viscosities at 100°C that fall in the middle of the SAE 90 range, and the fifth – the OlioFiat product – falling at the bottom of the range. None of the oils contains a VI improver, and they are all based on mineral oil. Given that the only recommended product that has its viscosity at the low end of the range is the Fiat brand – and Fiat made the car – it's safe to conclude that this type of product is acceptable. I could safely choose a non-EP SAE 80W/90 gear oil from any reputable supplier. However, as was concluded in the previous section, it must be of GL-3 performance level, not GL-4/GL-5 (that is, it must be non-EP).

The drain interval for the Lancia's gearbox/transaxle is 18,000 miles (30,000 kilometres) and it holds only a couple of litres of oil. As cost is not really an issue with this small usage rate, I choose to use a synthetic oil. I could choose an SAE 75W/90 gear oil – again, non-EP – but I actually use a synthetic SAE 20W/50 engine oil. It's cheaper than the gear oil – which would also be synthetic – and, as you will see, it also suits another of my cars.

The manufacturer of this synthetic engine oil specifically recommends it for gearboxes. It is of GL-3 performance level. Although it's an SAE 20W/50 multigrade oil, it contains no VI improvers which could shear. Being synthetic, it has a high inherent VI. However, the clincher for me is that, being an SAE 50 engine oil, its viscosity will definitely fall in the middle of the 'old' SAE 90 gear oil range (see Figure 8.15). When you buy an SAE 75W/90 you'll get an oil that falls in the 'new' SAE 90 range, and it will be lower in viscosity in the operating range up to about 125°C (see Figure 8.16).

Turning now to my 1969 Porsche 912, the Porsche specification is very succinct – use an SAE 90 hypoid gear oil. In 1969, 'multigrade' gear oils did exist – more widely in the US than in Europe – but Porsche wasn't recommending their use.

Multigrade engine oils and gearboxes

From the perspective of shearing of the lubricant, a gearbox is a far more stressful environment than an engine. If you wish to take advantage of some of the benefits that a multigrade engine oil can give in certain gearboxes then you must use an oil which is essentially completely shear stable – one with a shear stability Index (SSI, Section 4.1.1) of zero, or very close to it.

Oils that are fully synthetic can have these characteristics and, fortunately, they are widely and increasingly available. However, with the current lack of precise definition of what constitutes a synthetic oil (see the panel near the beginning of Section 8.1.2), you cannot simply make an assumption. To ensure that the oil you are considering has the shear stability level that you require, you will have to obtain this information (the shear stability index) from its marketer or manufacturer. Do not presume that a product described as synthetic will be shear stable.

Which Oil?

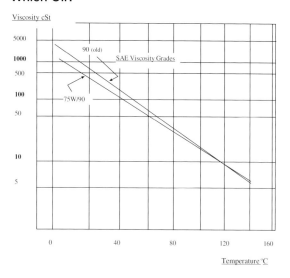

Figure 8.16 SAE 90 gear oil vs multigrade SAE 75W/90.

Porsche wanted you to use an oil that fell in the middle of the SAE 90 range. Obviously, it was using the SAE viscosity scale that was current at the time – the 'old' scale in Figure 8.15.

The conclusion with respect to viscosity for my Porsche is thus exactly the same as for my Lancia – SAE 90, as specified under the old scale. However, the Porsche requires GL-5 (hypoid) performance, and so I cannot extend my rationalisation and use the synthetic SAE 20W/50 engine oil. Instead, I choose to use an SAE 75W/90 synthetic hypoid diff oil (non-limited-slip). Its viscosity at 100°C is lower than it would have been in 1969 but, being a genuine multigrade, it is at least as viscous as the recommended oil at higher operating temperatures (see Figure 8.16.

For my pattern of driving, which is not stressful on the transmission, I consider this product to be satisfactory. If I were involved in motorsport, I most certainly would reconsider my decision. I would probably switch to an SAE 75W/110 product, provided I could find one on the market. SAE 110 is one of the new grades

---

Limited-slip differentials

In general, the limited-slip differentials that are found in classic cars contain mechanical clutches. They operate wet, within the gear oil.

As is indicated in Figure 6.4, the coefficient of friction between sliding metal surfaces in an oil usually increases as the relative sliding speed between the two surfaces approaches zero. Under such a circumstance, it is difficult to get the two surfaces to come to rest smoothly. They are very prone to the phenomenon known as 'stick-slip.' Stick-slip manifests itself as shuddering.

By contrast, if the friction coefficient reduces as the relative speed approaches zero, then the slow down will be inherently smooth. Such a situation can be arranged by adding the correct friction modifier (Figure 6.4). For this reason, manufacturers of limited-slip differentials recommend that they be lubricated with special limited-slip gear oils. As well as the appropriate EP additives, these oils contain friction modifiers. If your car has such a differential, then you should follow the manufacturer's requirements. If you don't, then not only will you probably experience shudder, but the clutches will also wear prematurely.

Choosing appropriate lubricants for your car

(see Figure 8.15), and has a viscosity that falls in the top half of the 'old' SAE 90 grade.

For the 1961 Simca Aronde, the choice is simple. The recommendation is viscosity grade SAE 90 (old scale) EP gear oil for both the gearbox and the differential. The situation is exactly the same as for my Porsche. A conventional SAE 80W/90 EP gear oil would suffice, or I might use an SAE 75W/90 synthetic if I wanted to give the Aronde a treat.

My other current car is the 1936 Austin Seven. Austin specified the following for my geographic area of use:

Gearbox: SAE 50 engine oil
Differential: SAE 140 extreme pressure gear oil

For the gearbox, I use the same SAE 20W/50 synthetic engine oil that I use in the gearbox/transaxle of my Lancia, and for the same reasons – better high temperature protection and better low temperature lubrication and gearshifts, plus full shear stability.

Turning to the Austin's rear end, SAE 140 is a high viscosity oil to use in such a light vehicle of very modest torque and power. EP additives did not come into general commercial use until the 1930s, and the Austin Seven's basic design predates this (although I acknowledge that the design of its differential did evolve over that period, although none of the designs involved hypoid gears). Austin's viscosity specification remained unchanged at SAE 140 from 1923 to 1936. I suspect that Austin found that SAE 140 oils gave better protection in the early cars. These cars and oils predated EP and anti-wear additives and, if you recall the Stribeck curve (Figure 3.5), increased viscosity helps to maintain full-film lubrication.

The downside of the high viscosity is the friction or churning loss, something that my Austin Seven – with its 747cc engine delivering well under 20 horsepower – can ill afford. Accordingly, I choose to use a synthetic SAE 80W/140 EP gear oil. Figure 8.17 shows how its viscosity compares with both a modern and a late-1920s SAE 140 monograde product (when base oils were less refined and had lower VIs). My choice gives better protection than the original oil when hot, and much better lubrication and lower power losses when cold. If companies should begin to market oils meeting the new SAE 110 viscosity classification I would consider one of those seriously.

Finally, there's the Rolls. Referring back to the previous section, the 'final' recommendation from R-R for the gearbox was to use an oil of viscosity SAE 90 (gear oil classification) or SAE 60 (engine oil classification). From Figure 8.15 – and using the old gear oil viscosity scale – we see that the desired viscosity is towards the top of the SAE 90 grade. However, as we have concluded that an engine oil is the preferred product, my choice today would be a synthetic SAE 20W/50 or 20W/60 engine oil, one with good anti-wear performance. A shear stable 25W/60 engine oil would also suffice.

Figure 8.18 shows the viscosity/temperature characteristics of an SAE 60 monograde oil typical of the 1930s, together with synthetic SAE 20W/50 and 20W/60 oils of today. At higher temperatures (above 100°C), the SAE 20W/60

Which Oil?

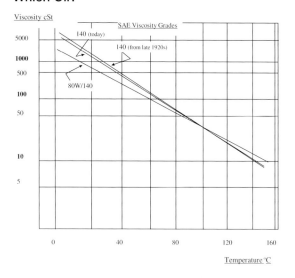

Figure 8.17 SAE 140 gear oils vs multigrade SAE 80W/140.

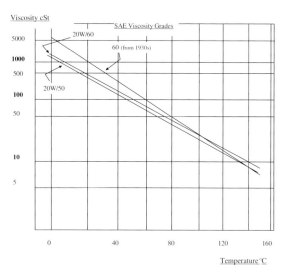

Figure 8.18 1930s SAE 60 vs modern multigrades.

multigrade oil will give better protection than an SAE 60 oil of the early 1930s. The SAE 20W/50 will also become superior as temperatures increase. At low temperatures, both oils will give much easier gear changes. If the car were to operate in a hotter climate, I would use the SAE 20W/60 (or 25W/60 if this is more readily available), while, in colder areas, I would choose the 20W/50. In either case, the multigrade engine oil must be synthetic to have adequate shear stability for the gearbox environment.

In the sequence of Rolls-Royce handbooks, over time there appears to be a trend of increasing the recommended gear oil viscosity. As stated above, in the final handbook, an SAE 60 engine oil is specified for the gearbox, whereas in the previous handbook, an SAE 40 engine oil was recommended (the Castrol XXL). I can give no explanation for this other than that the viscosity of engine oils produced was perhaps becoming more controlled. An alternative explanation is that Rolls-Royce's customers were experiencing some wear.

Turning to the rear axle, the R-R recommendation is for an 80/90 viscosity version of the same oil used in the gearbox. This is, at first sight, another somewhat

Choosing appropriate lubricants for your car

puzzling recommendation. In 1930, the SAE 80 gear oil viscosity grade classification specified only the viscosity of the oil at low temperature. It was the precursor to the 80W grade, the W terminology being introduced some years later to make this clear. Hence Rolls seem to be recommending an oil of the same viscosity as used in the gearbox (SAE 90), but with particular concern that it be mobile at low temperatures. Why the same concern did not exist for the gearbox oil is a puzzle. It may be that, by 1930, quality engine oils (as were recommended for the gearbox) invariably contained Pour Point depressants (see Table 5.1) but gear oils did not. However, this is not a very convincing explanation because at least one of the products recommended for the gearbox – the BP Energol 90 – was not an engine oil.

My choice for both the gearbox and the rear axle of the Rolls when operating in Sydney would be an SAE 20W/60 synthetic engine oil of GL-3 anti-wear performance level.

8.2.3 Automatic transmissions

If you have not already read Section 6.2 you would certainly benefit from doing so now. It will help you with the terminology used in this section. Also, it explains that I am dealing only with 'conventional' automatic transmissions. These are the units that (usually) consist of an epicyclic gearbox, a torque converter (or fluid coupling), and a series of wet clutches and bands.

The primary information that you need in order to chose an appropriate lubricant for your automatic transmission is whether or not it requires a fluid that contains a friction modifier additive. Unfortunately, it's most unlikely that the original manufacturer will have stated the lubricant requirements in such terms. They probably referred either to the name of a particular specification of fluid (eg DEXRON) or they listed particular products, hopefully from several brands or suppliers.

Of the US manufacturers, the simple rule that applies in the vast majority of cases is that, if your car was built prior to about 1974, it will most-likely require a non-friction-modified fluid if it has a Ford-sourced transmission, and a friction-modified fluid if it's from any other manufacturer. However, there is one major and one minor exception to this rule of thumb. The minor one is that if you have one of the relatively rare automatic Fords built prior to 1961, it will probably require a friction-modified fluid. It would have used GM Type A or Type A Suffix A originally.

The major exception is early Borg-Warner automatic transmissions. As was mentioned in Section 6.2, the first Ford automatic transmission was actually a Borg-Warner design. With this close connection, it's no surprise to find that, for early Borg-Warner transmissions that were used in other vehicles a little later (usually branded as Borg-Warner units), the fluid specification is non-friction-modified. A prime example of such a transmission is the Borg-Warner T35 which is found in a variety of cars in the US, Europe and other parts of the world. Check the original vehicle handbook recommendation if possible. If it specifies Type A or DEXRON, it requires friction modification. If it specifies a Ford fluid (eg M2C33 plus a letter in the range D to G), it requires a fluid with no friction modification.

Which Oil?

For the European manufacturers, the situation is less clear cut. With the smaller cars and higher price of fuel in Europe in comparison to the US, automatic transmissions were much slower in being accepted, and, to this day, they remain a minority. In the early days their presence was limited to larger luxury vehicles. Because of the low numbers of units required there was little incentive for manufacturers to develop their own units. GM or Borg-Warner transmissions were often used, but, as the market has increased, there has been a transition over time to in-house designs.

For some European cars – as explained in Section 6.2 – the manufacturers recommended Type A Suffix A fluid. This is the General Motors specification that preceded the introduction of DEXRON in 1967, but you'll find some European manufacturers still recommending it well into the 1980s. The reason for this was the concerns over DEXRON IID that arose with its introduction in 1975 (again, see Section 6.2). Some European manufacturers who had moved their recommendations to DEXRON reverted back to Type A Suffix A. A prime example was Mercedes (Daimler-Benz).

All of the automatic boxes that were involved in this specification rethink require friction-modified fluids. In reality, the concerns over DEXRON IID were unfounded, and, if you have such a transmission, you can safely use a DEXRON fluid.

Finally, if you have a Japanese car with an automatic transmission of Japanese manufacture then it almost certainly requires a friction-modified fluid.

Today, all modern automatic transmission fluids contain friction modifiers (they are needed for the lock-up torque converters that are now the norm). It is generally safe to presume that, if your automatic transmission was manufactured post-1974, it will require a friction-modified fluid. If you have a vehicle that requires a non-friction-modified fluid you will have to seek one out at your local oil supplier. Such fluids are still readily available (usually quoting Ford Type F or M2C33-F on the label, sometimes M2C33-G or Type G in Europe).

If your vehicle requires a friction-modified lubricant then all of the current fluids are potentially suitable. There is a proliferation of products on offer because different modern manufacturers may require slightly different friction, low temperature, or service-life characteristics. However, the variations between them are small, and they all will be suitable for your vehicle. However, if your car predates about 1985 – and you do not live in a cold area – it may be sensible to seek out an older DEXRON specification fluid. DEXRON IID fluids are still available. Their viscosity at low temperature is slightly higher and they will not be quite as prone to leaking out of your box as the modern versions are.

As a generality, using a friction-modified fluid in an automatic transmission that specifies a non-friction-modified one is potentially more problematical than the reverse situation. Overheating and wear of the friction surfaces may result. By contrast, the use of a non-friction-modified fluid in a box that specifies friction-modification will generally result in more abrupt shifts but not in rapid wear. However, in both cases, it's far wiser to use the correct fluid.

Choosing appropriate lubricants for your car

8.2.4 Overdrive units
In classic cars, these units usually contain an epicyclic gear set, brought into operation by a brake band or cone clutch. These, in turn, are either hydraulically or mechanically operated, the operation being triggered by a solenoid valve. The bands and clutches operate wet, within the gear lubricant.

The lubrication needs are similar to an automatic transmission. A level of anti-wear performance is needed but not full EP. Also, the units tend to contain 'yellow' metals, such as bronze, and EP additives would be corrosive. Most manufacturers of the overdrive units found in classic cars recommended monograde engine oils, typically SAE 30 or 40, when they were new. These products contained anti-wear additives and were readily available. Automatic transmission fluids have been recommended, but these have a lower viscosity – equivalent to an SAE 20 engine oil – and wear and leakage have at times been problems.

Historically, multigrade engine oils have not been recommended. Traditional multigrade oils were not sufficiently shear stable. However, today, the use of a shear stable synthetic multigrade engine oil is quite appropriate and can be beneficial. If the original viscosity recommendation was SAE 30, then SAE 5W/30 would be a good choice, while, if the original was SAE 40, a possible choice would be SAE 10W/40.

It's important that the oil chosen is shear stable. If you're unsure, use a monograde. However, irrespective of the viscosity grade selected, it is essential that the oil used has good anti-wear capability. I would avoid the modern low phosphorus oils (API SM, SN in the US or the C Sequence oils in Europe). For more information on this, see Section 9.1.3).

8.3 Chassis and wheel bearings
If your car has drum brakes then, in general, you'll be able to use a single grease for all chassis, steering and wheel bearing lubrication needs. The safest, most cost-effective grease type to use is non-complex lithium multipurpose grease (sometimes referred to as lithium stearate or lithium hydroxystearate grease). Check the vehicle manufacturer's original recommendation, but, again, as a fairly safe generalisation, a grease consistency of NLGI 2 will be appropriate.

If your car has disc brakes, and these are located adjacent to the hubs, then a conventional lithium grease will not be adequate for lubrication of the wheel bearings in those hubs. Its melting point will not be high enough. You'll need a so-called complex grease, or one made using special thickeners – clay (bentone), polyurea, etc. An NLGI 2 consistency is, again, probably the most appropriate.

In general, you will be able to use the higher temperature wheel bearing grease for the chassis and steering applications also. However, you must be alert to the possible problem of incompatibility with any existing grease. Clay greases are particularly sensitive in this respect. If using one, you must clean out all the old grease first (unless you know with certainty that it was also a clay grease).

If you wish, you can save money by confining the use of the high temperature grease to those wheel bearings where it is relevant, and using a conventional multipurpose lithium grease for all other locations.

Which Oil?

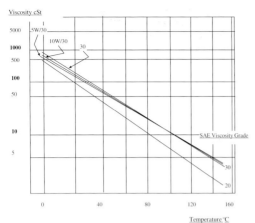

Figure 8.19(30) Viscosity vs temperature for SAE 30 engine oils.

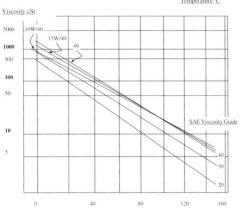

Figure 8.19(40) Viscosity vs temperature for SAE 40 engine oils.

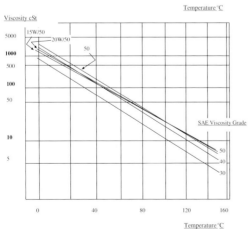

Figure 8.19(50) Viscosity vs temperature for SAE 50 engine oils.

9. Answers to some common questions

Some questions regarding the choice of lubricants for classic cars come up so regularly in club forums or in threads on the internet – and manage to generate such a range of answers – that I thought it might be useful to list them and give my own answers. Obviously, the matters that I have chosen to deal with are not the full gamut of such questions, but they are common ones.

9.1 Engine lubrication
9.1.1 My car has no oil filter (or a partial filter system). Should I avoid engine oils that contain detergents and/or dispersants?

In general, if your car has no oil filter (or only a partial filter system), you will be better off using an oil with detergents/dispersants than one without them. I am very aware that this is the opposite of the answer usually given in internet chat forums. A couple of typical statements are reproduced below:

"My understanding is non-detergent oils should be used in combustion engines that have no oil filters. By using non-detergent all of the crud goes to the bottom and is removed when the oil is changed. When a detergent oil is used the particles are kept in suspension and circulated around the motor. This is a good thing if you have an in-line filter but causes premature wear when there is none."
"What I have been told is, if it has a filter use detergent, if it does not have a filter use non-detergent."

The idea that, if you do not have a filter, solids formed from inadequate combustion or other sources, will sink to the bottom of the sump is a plausible but naive one. At the time that they are formed, combustion solids – and these are the prime source of solid particles – are microscopically small. They will not settle out under gravity. They will either adhere to metal surfaces or, over time, to each other. Eventually they will grow big enough to be seen and to settle.

While they are still microscopic, solid particles do very little damage. Hence it is best to keep them at this size if possible. This is the function of oil dispersant additives. Together with detergents, they will also discourage the tendency to adhere to metal surfaces and to accumulate there. Also, if particles are allowed to grow they will reach a size where they do damage before they are large enough to settle out.

An exception to this advice is if your car is a vintage model and burns large amounts of oil – as an inherent part of its lubrication system design – and requires frequent decarbonising. Detergents, and to a lesser extent, dispersants, may increase the quantity of material to be decarbonised (see the answer to 9.1.4 following also).

Which Oil?

There also tends to be an automatic presumption that full flow filtration is better than partial. To a significant extent this is true. That is why the technological trend with engine design over the years has been from no oil filter, to partial oil filter, to full oil filter. However, both partial and full systems have compromises.

The filter in a full system has to be coarser than that in a partial. With all of the oil going through the filter there is a limit to how small the filter pores can be. If they were too small the pressure drop (resistance to oil flow) would be too high. In a bypass system, a finer filter can be used as its impact on oil flow to the engine is less.

This is shown clearly by the filter systems used, for example, on large diesel engines, such as in many Cummins trucks. Cummins used both a full flow and a partial flow filter. The full flow filter has a nominal 30 microns mesh, while the partial flow filter has 10 microns.

Be aware that a mesh of, say, 10 microns, will not automatically filter out all particles with a size greater than 10 microns. All filters operate at less than 100 per cent efficiency. There is an SAE test (J1858) that measures this. Typical results for the types of paper element filters sold for use in car engines are 40 per cent capture efficiency at 10 microns, 60 per cent at 20 microns, 93 per cent at 30 microns, and 97 per cent at 40 microns. For comparison, engine bearing clearances typically range from 15 to 50 microns (0.0006 to 0.002 inches).

If the only source of solid particles in your engine oil were the combustion

---

A common confusion

One quite common confusion involves the viscosity rating of engine oils and their detergent content. Here is a statement from a web forum:

"Non-detergent oil, such as SAE 30, is not used in modern passenger vehicle engines. It is still used in some gasoline engines such as lawn mowers."

There is no necessary connection between the viscosity grade of an engine oil – defined by its SAE grade – and its detergent content. The SAE 30 oil used in lawn mowers, that is mentioned above, certainly contains detergents. Detergent content relates to the engine performance specification of the oil, not the viscosity. Under the US system, this performance specification is given by the 'S' rating. A non-detergent oil will have a rating of SA or SB.

The confusion probably arises because, back when engine oils did not contain detergents – that is, prior to at least 1963 but generally much earlier – there were also no multigrade viscosities. All engine oils were monogrades (eg SAE 30). Multigrade viscosities and the universal use of detergents developed at around the same time, during the 1960s. Hence the assumption by some people that they are linked, and the further assumption that, because monograde oils before 1960 (or a little earlier) did not contain detergents, this is still the case today. There is no necessary link. A non-detergent oil can be either mono or multigrade, as can a detergent oil. The choice is entirely in the hands of the blender of the oil.

Answers to some common questions

process, then you would actually do better to use a partial filtration system and a finer filter – one with an efficiency in the high 90 per cent at 10 microns – in combination with a detergent/dispersant oil. The dispersant would limit the growth of particles and, any that did grow, would be removed by the filter before they caused significant damage. The fact that particles might go through the engine quite a number of times before they encountered the filter would not be a problem because of their size.

However, combustion is not the only source of particles. There is engine wear and there is dirt. Wear particles are typically also very small and so the same arguments apply to them as I have just given for combustion particles. By contrast, dirt particles are very difficult to avoid (air filters are also not perfect) and they can be of a size that will cause immediate wear. Full filtration is the best defence against them.

Hence, the ideal system would be to use both a full and a partial filter system. This is what is done on large trucks, and there are some systems sold by aftermarket suppliers for cars. However, to return to the original question, irrespective of the type and efficiency of your car's oil filtration system, it is best to use a detergent/dispersant oil. Relying on gravity to remove solid particles will almost certainly ensure that they have grown to a size where they cause damage before they finally are heavy enough to settle or be trapped by a filter.

Irrespective of which filtration system (if any) your car has, you should ensure that you drain the engine oil regularly. Do not run extended drain intervals, measured either in terms of distance travelled or operating hours.

9.1.2 The original specification for my engine was a monograde oil. Can I use a multigrade?

The answer is in general "Yes," but with a few caveats.

Once again, the first one applies if you have an engine that inherently burns large quantities of oil. You probably will already have a problem with carbon deposits in the combustion chamber(s) and on piston crown(s). A conventional, non-synthetic multigrade oil – one that generates its multigrade characteristic by the use of a polymeric viscosity index improver – will tend to increase the amount of carbon. A fully synthetic oil (with inherent multigrade characteristics) may reduce deposits but it will cost you a lot to keep up the supply of oil to your engine. This sort of high-consumption problem is generally confined to pre-1930 vehicles.

The second caveat depends on the viscosity grade and the shear stability of the multigrade oil that you use. Suppose that the original specification was for an oil with a viscosity of SAE 40, and you choose to use a low shear stability 15W/40. In service, the oil will shear down to at least an SAE 15W/30 grade and possibly lower.

Oil pumps in car engines are usually of the positive displacement type. The oil flow rate that they deliver depends mainly on the speed at which they operate. With a lower oil viscosity you may still have adequate oil flow but the lower viscosity will mean that the pump generates less pressure for the same flow rate. You will observe a lower oil pressure – if you have a gauge – in the lower rev range. At higher

Which Oil?

revs, where the pressure may reach the relief valve setting, you may not notice any difference. At idle you may see a significantly lower oil pressure.

This outcome is not of fundamental concern provided that the viscosity is still high enough to do the lubrication job. However, if you have an engine that has a worn oil pump, or an engine where the original pump has relatively large design clearances, then you may have a problem.

The oil on the discharge side of a pump is at a higher pressure than the oil on the inlet side (this is one description of the function of the pump – to generate an increase in oil pressure). However, if clearances within the pump are large, some oil will flow from the discharge side, back to the inlet side. This flow will occur within the pump body and it tends to be greater if the viscosity is lower.

At high engine speeds – which means high pump speeds – this internal oil flow will be a modest percentage of the total oil flow delivered to the discharge side of the pump. However, at low engine or pump speeds – such as at idle – it may be a significant percentage. In this circumstance, unlike the situation describe above, the oil flow to the components of the engine, downstream of the pump, may drop below the minimum design quantity. Lubrication may be inadequate.

You should note that the oil pressure behaviour observed will depend on the location of the oil pressure sensor. If it's immediately at the pump discharge it may not drop very much. However, if it's well downstream of the pump you may see a quite large drop. If you do not have an oil pressure gauge you will not see these effects. However, if the oil pressure light comes on (or flickers) at idle when your engine is warm, you should be concerned.

Overall, the message is that it's fine to use a multigrade oil provided that it has good shear stability. Use a reputable brand, and do not buy on lowest price.

9.1.3 Can I use modern oils which have low levels of zinc in my older engine?

This topic, or a variation of it, is probably the hottest one in classic car circles at the moment, particularly in the US. A couple of examples of discussions on the internet follow:

"In recent years, auto manufacturers have switched to roller camshafts for performance and fuel economy reasons. This, combined with the EPA requirements for reduced emissions, has led to the reduction of zinc dialkyl dithiophosphate (ZDDP). This lower zinc content will not properly lubricate the flat tappet hydraulic lifters we use on the majority of our engines."

"I've been hearing about way too many camshafts going flat nowadays – particularly on freshly built motors using flat tappets."

"Zinc has been added to oil for years as an anti-wear additive – particularly crucial where there is metal to metal contact in the engine – ESPECIALLY flat tappet to camshaft contact."

"Today, flat tappets don't exist in most new cars. Today, emissions have pressured oil companies to remove zinc from oil. So, a necessary component for older cars has been going away … BEWARE."

Answers to some common questions

"BOTTOM LINE: Running an old engine without zinc in the oil is bad news – especially when breaking in a fresh build. Ask around and you'll hear about high rates of camshafts going flat. Avoid the problem by choosing the right oil – one that has a zinc content of at least 0.11 per cent. If you're running zero zinc content oil in an old engine, I believe you're asking for trouble."

As is very often the case, these internet posts contain a mix of fact and fiction. Some of the truth is straightforward and undisputed. The US specifications for gasoline engine oils have put an upper limit on phosphorus content since 1989. The reason for this is to preserve the performance of the catalytic converter in the exhaust or tailpipe. Phosphorus in the exhaust gases from the engine poisons the catalyst (see Section 4.5).

The maximum amount of phosphorus allowed in an engine oil was set at 0.12 per cent with the introduction of the API SG engine oil classification in 1989, although many oil blenders – in the US and elsewhere – had adopted this limit voluntarily well before then. There was a further reduction to 0.10 per cent in 1996 with the introduction of SJ and a reduction to 0.08 per cent in 2004 with SM.

In Europe, up until 2004 when the 'C' sequence or catalyst-compatible oils were introduced as a separate specification by ACEA (see Section 5.1.2), oil specifications broadly followed the US lead. However, since 2004, the very low phosphorus oils have largely been confined to the C sequence products. There are no phosphorus limits for the A/B sequence oils, although the level of phosphorus has to be reported. In general, they run at 0.10-0.12 per cent.

Why am I talking about phosphorus when the issue of concern is apparently zinc? As explained in Sections 4.3 and 4.5, phosphorus is added to engine oils as part of a chemical called zinc dithiophosphate (ZDTP). A restriction on the amount of phosphorus led automatically to a restriction on ZDTP and thus to a restriction on zinc.

ZDTP is added to an oil for two reasons. It acts as an antioxidant, and also gives wear protection under mixed lubrication conditions. In automobile engines, sliding cam followers operate in a mixed or partial lubrication regime, and so anti-wear performance is essential in their lubricants. It's the reduction in ZDTP that has given rise to the concerns over potential camshaft/follower wear in older cars. Zinc level in the oil is being used as a shorthand way of expressing this concern.

Unfortunately, in this case, some very important information is being lost by this use of shorthand. Zinc dithiophosphate (ZDTP) is a complex chemical compound. In reality, its name is already shorthand. It stands for two families of compounds,

---

Zinc and anti-wear

While the majority of the wear debates on the internet focus on zinc levels in the engine oils, it is important to realise that zinc is not the anti-wear chemical. In the molecule zinc dithiophosphate (ZDTP), it is the phosphorus and the sulphur (the 'thio' portion of the molecule) that are the key to anti-wear performance. Do not imagine that adding zinc to an engine oil will give it anti-wear properties.

Which Oil?

one called the zinc di-alkyl dithiophosphates, and the other the zinc di-aryl dithiophosphates (and there are mixed families, the zinc alkyl-aryl dithiophosphates). And we are not finished yet! The words alkyl and aryl are, in themselves, shorthand, and they stand for a whole host of possible organic chemical groups (of two particular types).

So, in talking anti-wear performance we cannot simply talk zinc. The performance has to be determined by engine testing. However, the second contributor to the net is absolutely correct in stating that running zero zinc in an old engine is asking for trouble. It would be asking for trouble to use such an oil in any engine built since about 1925. Fortunately, no such oils exist in the general marketplace. To find one you would have to go to a specialist supplier and ask for an engine oil of SA performance level.

As the formulators of engine oils have been forced to reduce the amount of phosphorus in their oils they have done at least two things. Their first move was to change to chemical versions of ZDTP that gave better anti-wear performance for the same amount of phosphorus (see Section 4.5). In this way they could, at a cost, reduce the quantity of additive necessary (and hence reduce the phosphorus level), but maintain the anti-wear performance.

However, in doing this, the antioxidant performance of the additive tended to be reduced. The formulators compensated for this by using alternative antioxidants that contain no phosphorus. They also do not contain zinc but, again, they are far more expensive than the ZDTPs used. This change in chemistry comfortably handled the mandated reduction in phosphorus down to 0.10 per cent.

The second and more recent move has been to introduce anti-wear compounds that do not contain phosphorus. There is a variety of approaches but the most common is to use compounds that contain an oil-soluble form of molybdenum.

A suggestion made widely on the internet is that a readily available solution is to use an oil that has been formulated primarily for heavy-duty (truck) diesel engines. These oils have not had to reduce their levels of ZDTP nearly as much as motor car oils because, at least until recently, the vehicles involved did not have exhaust catalysts and phosphorus poisoning was not an issue. However, this too is changing as this post shows:

"This bottle of XXX-brand oil was rated CJ-4 which is the latest (as of 2007) specification for diesel oils and follows the spark engine oil trend of low zinc content. Only diesel oils rated CI-4 or earlier (CH-4, CG-4, etc) have the proper zinc content. As for spark engine oils, look for an API rating of SE or earlier (SD, SC, etc) for the proper zinc content. Don't do what I did and buy diesel oil expecting it to have enough zinc for your engine break-in."

Unfortunately, this advice, if followed, could lead to disaster. First, is it appropriate to use heavy-duty (truck) diesel oils in passenger car engines? If the oils for passenger cars and heavy-duty diesels were, in fact, readily and simply interchangeable, the oil manufacturers would not bother to develop, test and

Answers to some common questions

market two product ranges. The reality is that heavy-duty diesel engines are very different beasts to passenger car gasoline engines. For example, conditions in the combustion chamber area are far hotter and far more severe in diesel engines. The chemical additives used in their lubricants are chosen to work in that sort of environment, and they will not function nearly as well in a gasoline engine environment. The additives have the same functions and descriptions – detergents, dispersants, anti-wear additives, etc – and, in general, they come from the same chemical families – but they are not identical substances.

This fact, unfortunately, applies to the anti-wear chemistry. The ZDTP used is designed to activate (and survive) at higher temperatures than those that generally prevail in gasoline engines. Just because a heavy-duty diesel engine oil contains a higher level of zinc than a passenger car engine oil, it does not follow that it will be more effective in preventing wear in a car engine. This was shown clearly in some of the engine testing done in the US as part of the development of the SM performance classification – the one that introduced the lowered limit of 0.08 per cent phosphorus. A proprietary heavy-duty diesel engine oil that had a high zinc and phosphorus content – well above 0.08 per cent – failed the main gasoline engine test for wear performance, a test that involves an engine with sliding or flat tappets.

The message is, do not use heavy-duty diesel engine oils in passenger cars other than in an emergency. If you do, there will be no immediate catastrophe. The oil will function adequately, but not optimally. However, over time, your engine will experience more wear, and deposits may build up on the metal surfaces within it.

It's important to understand that these comments apply only to the diesel engine oils specifically designed and sold for use in large truck engines. Oils designed for passenger car and small truck/van diesel engines are, in general, acceptable. They will be labelled with a gasoline engine performance classification first and a diesel classification second – for example, SL/CF in the US or A3/B3 in Europe.

Turning now to the other advice given in the internet post – to use an oil of SE or earlier rating to get the proper zinc level – the reality is that, back in the 1960s, when the SD and SC oils mentioned were current, zinc levels were typically 0.08 per cent or lower. Phosphorus levels were a little below the zinc values. Hence, these oils had less ZDTP in them than the current SM and SJ oils that have caused the angst, and the type of ZDTP used was less active. Earlier oils (SA and SB) had even less ZDTP, none in the case of the SA oils (see Table 5.2).

So again, this is potentially dangerous advice. At the very least, if you're tempted to go back to the past, you should never use an engine oil with a performance level less than the manufacturer of your engine specified. There is a great deal more involved in the situation than the zinc level.

Many of the problems reported on the internet regarding cam or follower wear – the camshafts going flat problem – involve engines that have had performance camshafts fitted. We should be cautious in extrapolating this experience to cars with their original cams or cam profiles. First, sports cams invariably involve higher lift and/or faster opening and closing profiles, plus stronger valve springs. All of these changes increase the loads on the contacting surfaces, sometimes by very large

Which Oil?

amounts. Remembering the Stribeck curve (Figure 4.6), the increased loads may be sufficient in themselves to move the lubrication situation from mixed to at least partially boundary, where a chemical anti-wear additive such as ZDTP may not be adequate.

Also, these cams are installed new. At initial start-up they do not have any protective chemical film on the rubbing surfaces, derived from in-service use. The running in process used is critical. It is not generally realised that the chemical reaction that allows ZDTP to form a protective film does not proceed at room temperature or even at modest temperatures. The exact temperature at which it begins varies with the chemistry of the ZDTP additive involved. However, running in an engine at too low a temperature or load can lead to excessive cam wear. This dependence of the reaction temperature on the chemistry of the ZDTP is one of the reasons that heavy-duty diesel engine oils may not give adequate anti-wear performance in a gasoline engine, even though they contain higher levels of zinc. The chemistry has been optimised to suit the conditions in diesel engines, not gasoline engines.

So, what is the answer to the question? Can I use a modern, low zinc engine oil in my classic car? For the vast majority of owners, the answer is yes. These modern oils have to pass stringent engine wear tests. Their anti-wear performance is certainly superior to the oils sold prior to the early 1970s.

For owners of cars that were produced later than the early 1970s, and that have the original specification camshaft, the questions is, how highly stressed was the original engine – is it a high-performance model? If it is highly stressed then I would not use a low phosphorus oil. I would make this same decision if my engine has been modified to give greater performance. For all other engines I would not have a concern.

Which oils can I use if my engine is one that should avoid the new, low phosphorus products? The answer is – for owners in the US or other regions where the US engine oil specifications predominate – an oil with an API engine performance classification of SL (or earlier if appropriate). The oils to be cautious of are those that meet the API SM and SN specifications.

Owners who live in Europe are much better catered for in this respect than the rest of us. As has been explained (see also Section 5.1.2), the European (ACEA) engine oil performance classification system has a separate category for catalyst-compatible oils. These oils have a performance classification that starts with C (C1, C2, etc). European owners should play it safe and avoid these products – unless the manufacturer of their engine specifies their use – and instead choose an appropriate product from the oils that have a performance classification that involves the letters A and B (A2/B2, A3/B3, etc). The ultra-low phosphorus oils are generally confined to the 'C' oils.

Oils blended to the European specifications are available in many markets outside of Europe. Read the labels on the cans or bottles, and look for the ACEA classification. These ACEA-qualified oils also have to meet an additional valve train wear test – additional to the US API qualification requirements.

Answers to some common questions

9.1.4 My engine is air-cooled. Should I avoid:
a) Oils that contain detergents?
b) Multigrade oils?

It is a fact that the combustion chambers and upper cylinder walls of air-cooled engines tend to run hotter than the same areas of water cooled engines. Hence, the behaviour of engine oils under these more severe conditions is very important. The critical issue is the behaviour of the film of oil that is deposited every cycle on the cylinder walls to lubricate the motion of the pistons. In turn, the critical performance factors for the oil are its viscosity in this high temperature/high shear zone, and its deposit-forming tendency. The latter is a function of its volatility, its oxidation stability, and its ash-forming tendency.

Detergent additives contain metals, and these are left behind as ash when an oil burns. Conventional multigrade oils contain viscosity index improvers. Viscosity index improvers are polymeric substances and they tend to not burn completely, leaving some residue. They also have a tendency to shear, lowering the viscosity of the oil. Another concern is that the base oil used in a conventional multigrade is lighter – and hence more volatile – than that used in a monograde of the same nominal SAE grade (see Section 4.1.1).

Given these facts, you'll hopefully agree that the questions posed in this section have a good basis. The higher temperatures that can exist in the combustion areas of air-cooled engines could exacerbate all of these potential problems (and they do).

It is, then, of no surprise to learn that, when multigrade oils first came on the scene in the early 1960s, VW and Porsche – the manufacturers that people tend to be talking about when they refer to air-cooled engines – did not endorse their use. These oils potentially could have increased any deposit problem and, being of questionable shear stability, also could have resulted in a lower viscosity and less lubricity in the vital piston ring/liner area.

However, the story is different today. Conventional multigrade oils – ones formulated from mineral oil plus a viscosity index improver – now contain balanced detergent/dispersant additives (and generally use viscosity index improvers that have inherent dispersant characteristics). They do not give increased deposits. The viscosity index improvers used are also far more shear stable (if you use a reputable brand of oil with a reasonable engine performance level).

Today, you also can choose synthetic oils. These can be multigrades, without requiring the addition of any polymer, and they are completely shear stable. The circle has completely turned and, for many years now, Porsche has strongly preferred synthetic, multigrade oils for its modern air-cooled engines. VW no longer makes such engines, but in countries such as Brazil, where the old VW Beetle design was produced under licence for many years after it had been dropped in Europe, multigrades became the manufacturer's recommendation for the engine.

9.1.5 Can I/should I use a synthetic engine oil in my classic?

As explained in Section 3.1, there is, unfortunately, no universal definition of what a synthetic engine oil is. For the world outside of the USA, it generally means an

Which Oil?

oil formulated using a base stock that has been made (synthesised) by chemical reaction of relatively simple organic molecules to make more complex molecules. In the US it can also mean oils formulated using conventional mineral oil base stocks, that have been subjected to significant additional chemical refining. Fortunately, for the purpose of answering the question posed, this potential difference in source or definition is not critically important.

The performance strength of synthetic oils is mainly in their greater ability to withstand high temperatures. Their natural area of use is in high-performance, highly-stressed engines, or in engines running for long periods at or near maximum output. However, they have some other valuable properties of more general relevance. Some synthetics have inherently high viscosity indexes. It is often possible to produce a multigrade oil without the need to add viscosity index improvers. This avoids the problem of loss of viscosity through shear in service that polymeric VI improvers have. Also, some synthetic base oils have lower volatility, reducing oil consumption if lighter viscosity grades are used.

The negatives associated with synthetics are that, as a generalisation, they have less ability to suspend solids and other contaminants. They also tend to shrink rubber-based oil seals (synthetic and non-synthetic). However, today, all synthetic oils sold by reputable companies have had components added to ensure that these weaknesses are overcome. However, when synthetic engine oils were first commercialised widely in the late 1980s and into the 1990s, not all products were fully balanced, and some problems emerged. These early problems are the main source of the doubt that still lingers over their use in classic cars.

Today, provided it is from a reputable marketer and meets all necessary performance and viscosity requirements, a synthetic oil can be safely used in your engine. However, if you are a typical classic car user, and operate your vehicle relatively infrequently, you will not necessarily see any benefit as a return on the greater cost over a conventional oil. If you have a high-performance classic, or use it in strenuous activities, a synthetic could be a wise choice.

9.2 Manual gearboxes
9.2.1 Can I/should I use an engine oil?

This question comes up repeatedly because many owners of classics from the 1950s and '60s are surprised/concerned to find that the original manufacturer's recommendation was to use an engine oil in the gearbox. Here is a typical thread from a website discussion.

Contributor 1: "The car is a '59 XXX. I'm working with a reprint of an aftermarket workshop manual, the original driver's manual and advice from a restoration shop. The reprint shows 50 weight motor oil, the driver's manual shows 90 weight hypoid gear oil (GL-4) and the restoration shop suggests 50 weight motor oil.

Contributor 2: "I was told by the individual who rebuilt my transmission – who does this sort of thing for a living – to use only 30W non-detergent oil."

Contributor 3: "Here is the story of my experience. I bought my 1958 XXX brand new. I always changed the oil in the gearbox/overdrive with straight 30 grade engine oil. At that time this was the recommendation. The gearbox in my car does not have a synchromesh first gear and at 14,000 miles (22,000km) I had to have the lower layshaft with its straight cut spur gear teeth, plus all the bearings, changed. Perhaps I was too rough or too young, but I felt it didn't last very long.

"After getting it repaired, I was extra careful never to take it up above 3000rpm in first gear, and I made sure that I was at a full stop before engaging first. The same problems occurred again at 42,000 miles. Before I put the car away in 1972 with 80,000 miles on it, the first gear and bearings in the gearbox were damaged again. All this time I used 30 weight engine oil as the lubricant for the gearbox."

The 'problem' raised by the first contributor is resolved very simply. If you look at Figure 6.1, you will note that, under the old gear oil viscosity classification scheme (which applied at the time), an SAE 50 engine oil and an SAE 90 gear oil have the same viscosity. What is apparently required by this gearbox is an SAE 90 gear oil of GL-4 performance level. An SAE 50 engine oil has both the required viscosity and just about the required level of gear performance (it is more GL-3/4 than GL-4). The gear performance is provided by the anti-wear additive that is added to protect cams and followers (see Section 6.1.2).

The problems experienced by the third contributor are also readily understood. Referring again to Figure 6.1, you will see that an SAE 30 engine oil has only about half the viscosity (at 100°C) that the manufacturer specifies (SAE 50). A lower viscosity sends you to the left on the Stribeck curve (Figure 4.6). When an SAE 50 engine oil (or SAE 90 gear oil) is used, the gears operate in the mixed lubrication region, and the chemical anti-wear additives are adequate to protect them. The use of an SAE 30 engine oil, with its lower viscosity, moves the lubrication situation into the boundary region under some circumstances and, in this region, an anti-wear additive is inadequate (Figure 4.6).

Contributor 2 throws in a couple of red herrings, presumably unintentionally. First, there is no such viscosity grade as SAE 30W (see Table 5.6). The highest 'W' grade is 25W. The contributor presumably meant SAE 30. Secondly, detergents do not have any detrimental effects in gearboxes (see 9.2.2 following).

9.2.2 Can I use an oil that contains detergents?

First, we need to be aware that detergents are present only in engine oils. However, their absence from gear oils is not because they would cause harm, but rather because they would not be of any benefit. Here is a web string that covers detergents, antifoam additives and, also, the engine oil vs gear oil dilemma again:

Webmaster: "One member claimed that the 'correct' type of oil to use in a YYY gearbox is 30W non-detergent engine oil. A few people expressed surprise, stating that they had been using 80W gear oil. However, this is definitely incorrect oil for YYYs. The factory clearly specified in the Owners Manuals that engine oil is correct.

Back to the original statement about "30W non-detergent engine oil" – some of

Which Oil?

us then wondered whether the "non-detergent" aspect is important. Could we use a normal detergent-type engine oil instead?

Contributor 1: "Following up the thread that was running a couple of days back about detergent vs non-detergent oils, I did some research, including contacting tech departments at oil companies A, B and C.

"First, Company C explained that the term 'detergent' does not mean the oil has washing-up liquid added. It refers to a property of the oil, namely an ability to keep insoluble particles (like sludge) in suspension (rather than deposited somewhere). Certain additives are used to give oil this property, but they're not what we think of as 'detergents.'

"Next, foaming or frothing is a potential problem with both detergent and non-detergent oils, so both have 'anti-foam' agents added. To quote Company A: 'The defoam is there to prevent oxygen build up which occurs from circulation … it is not present to reduce the effects of detergency. Too much oxygen in any lubricating system will reduce the lubricating effects of the oil causing premature wear.'

"So it seems that potential foaming is not the reason for not using detergent oil in trannys. What then is? Well, Company B ignored the question. Company C thought it might be … '… because back in the '60s it was considered best not to emulsify any water that got into the transmission.' Company A said: 'There are several reasons why non-detergent oils are recommended, the main reason is that non-detergent oils are also very low in other additives such as zinc and calcium. Zinc and calcium can cause premature wear of certain components which may have been used in the gearbox.'"

Presuming that Contributor 1 correctly recorded the advice from the two oil companies that responded, then I can only sympathise with him. Some of it is bizarre.

Dealing with the easy bit first – the confusion over SAE 80W gear oil vs SAE 30 engine oil. Referring to Figure 6.1, you will see that these oils have the same viscosity. If the factory specified an SAE 30 engine oil for the gearbox, then a GL-3 performance level gear oil of SAE 80W viscosity would be perfectly suitable also. A GL-5 level product – and perhaps a GL-4 one – would not be suitable (see later in this answer also).

Now for the question of detergents. I should note that, although this question is invariably couched in terms of detergents, modern oils contain a mix of detergents and dispersants, and so this question is really, can I use an oil that contains detergents and dispersants? As Contributor 1 says, detergents/dispersants surround fine particles and stop them agglomerating. This, and the nature of the detergent/dispersant molecules themselves, helps to keep such particles in suspension in the oil and reduces their tendency to settle out.

In a gearbox, unlike an engine, the only particles are wear particles. If these are of any significant size a detergent/dispersant will not be able to stop them settling. If they are extremely fine, they will have a slightly greater tendency to stay in suspension. However, extremely fine particles are not the source of concern

Answers to some common questions

when it comes to wear. It is the larger ones. The presence of detergent/dispersants will make an insignificant difference. Hence the willingness of manufacturers to specify the use of engine oils where their viscosity and anti-wear characteristics are appropriate.

Contributor 1 also mentions foaming tendency. Again, the presence of detergents/dispersants may make a fractional difference to the foaming tendency of an oil. Air is readily entrained into the oil in a gearbox because its lubrication mechanism relies on splash. However, foam is a potential problem with just base oil alone. The use of antifoam additives is essential, irrespective of the presence of detergents/dispersants. This is shown by the time sequence that is listed in Table 5.1. The use of antifoam additives predated detergent/dispersants by 40 years. All engine oils contain antifoam additives, as do all gearbox oils. This is true regardless of the presence or otherwise of detergent/dispersants.

Similar comments apply to the concern raised by Oil Company C regarding the emulsification of water. It is true that detergents will increase this tendency, but it is of no real relevance to gearboxes. If water is getting into your gearbox then you will be in trouble irrespective of the presence or otherwise of detergents.

Finally, the statement by Oil Company A that " ... the main reason to use non-detergent oils is because they are also very low in other additives such as zinc and calcium ..." and that "... zinc and calcium can cause premature wear of certain components which may have been used in the gearbox ..." is, as I have said, bizarre. The only engine oils that can, in fact, be considered for use in a gearbox are those that contain adequate anti-wear additives, and these additives, at least historically, were zinc based (ZDTP). Rather than causing premature wear, zinc retards it.

The advisor is presumably confusing anti-wear additives with EP additives. EP additives, again historically, were almost invariably based on active sulphur, and this can corrode some metals commonly used in gearboxes (see Section 6.1.2). GL-5 performance level gear oils are what commonly have to be avoided, not engine oils that contain zinc. GL-4 oils contain about half of the level of active additive that GL-5 ones do, but they still need to be considered with caution.

In summary, oil manufacturers do not put detergent/dispersant additives into gear oils for the simple reason that they cost additional money, and, unlike in engines, they bring no benefits. They do not omit them because of a performance problem.

9.2.3 Can I use a multigrade engine oil?

The answer to this question used to be simple. It was "No." All multigrade engine oils used to contain polymers called viscosity index improvers. The types of polymers used were not sufficiently shear stable to survive intact in a gearbox. The viscosity of the oil would drop rapidly in service before stabilising. As the user, you would not have any way of predicting just what value the viscosity would stabilise at. It will depend on the shear stability of the polymer used, and this can (and almost certainly will) vary brand-to-brand. An example of the confusion caused is shown in this extract from another string on the web.

Which Oil?

"I looked for guidance on this subject about a month ago and searched both this forum and the 'net.' There are lots of opinions.

"I found it boiled down to SAE 30 or 40, in accordance with the original handbook, or good quality SAE 20W/50.

"Prior to reading this post I had come down in favour of the 20/50 because there were a number of very firm recommendations in favour of the 20/50 on various web sites. Now I am back in the ranks of the undecided."

You will note that the owners of the cars in question have, by trial and perhaps costly error over the years, reached the point where, if they use a multigrade engine oil, they use one that is more viscous than recommended originally by the manufacturer. They are relying on it dropping in viscosity in service, down to the specified range. Once the polymer has sheared it may very well be that the oil is no longer a multigrade. Very little may have been gained, and a risk has been run because the owners cannot be certain that the ultimate viscosity will be the required one.

An alternative approach is taken by Contributor 3 who was quoted in Section 9.2.1, the guy who had all the problems when using an SAE 30 engine oil when an SAE 50 was specified. In his string he went on to say that, following advice from a marque specialist, he now uses SAE 20W/50 engine oil and is very satisfied. So, why has this overcome his problem? Why doesn't the SAE 20W/50 simply shear down towards the problematic SAE 30 viscosity level? The answer is that it probably would and maybe it does. However, he doesn't give it time to cause a problem. He also tells us that he changes the gear oil every 3000 miles.

Fortunately, today, all of these owners can have the benefits of a multigrade oil – lower viscosity and better shift characteristics when cold, but greater gear protection when hot – without taking any risks. They can choose a synthetic engine oil which will be a multigrade but will also be completely shear stable. A good choice for the first group of owners (the ones who need to match an original specification of SAE 30 or 40) would be an SAE 10W/40 (or 5W/30 if in a very cold climate). SAE 15W/50 or 20W/50 would be a good choice for the second group, the ones who are working to an original specification of SAE 50 engine oil or SAE 90 GL-3/4 gear oil. However, in both cases, the oil must not be a conventional multgrade with a polymeric VI improver. It must be a shear stable synthetic and it must have good anti-wear performance.

10. Glossary

AAMA	American Automobile Manufacturers Association
ACEA	Association des Constructeurs Europeens d'Automobiles
API	American Petroleum Institute
ATF	Automatic Transmission Fluid
CCMC	Comité des Constructeurs du Marché Commun
CCS	Cold cranking simulator
CEC	Co-ordinating European Council
EP	Extreme pressure
ILSAC	International Lubricant Standardisation and Approval Association
JAMA	Japanese Automobile Manufacturers Association
MRV	Mini rotary viscometer
NLGI	National Lubricating Grease Institute
SAE	Society of Automotive Engineers
SSU	Saybolt Universal Seconds
SUS	Saybolt Universal Seconds
VI	Viscosity index
VII	Viscosity index improver
ZDTP	Zinc dithiophosphate

Index

AAMA 55
ACEA engine oil classifications 57-60, 118
Anticorrosion additives 35
Antifoam additives 34, 122
Antiwear additives 46 (see also ZDTP)
API engine oil classifications 13, 52-55
API gear oil classifications 66, 67
Automatic transmissions 72-75, 107, 108
ATF (automatic transmission fluid) 73-75
   choosing 107, 108

Base oils 23
Base oil groups 41
Boundary lubrication 37-39, 47

Catalytic converters 15, 17, 49
CCMC engine oil classifications 16, 57
Chassis lubrication 76-80, 109
Corrosion inhibitor 36

Detergents 35, 37, 111, 119, 121
Dispersants 35, 37, 111

Emissions 49
Engine oil
   choosing 81-97
   drain interval 45, 83-85
   Europe 56-60
   Japan 60
   low temperature properties 33, 62
   multigrade (see Multigrade viscosity)
   service classifications 52-55, 57-60
   synthetic 18, 62, 84, 85, 119
   US 52-56
   viscosity (see Viscosity)
   zinc level (see Zinc)

Extreme pressure (EP) additives 47-49, 66-68, 98

Friction modification 47, 51
   automatic transmission fluids 51, 73-75, 107, 108
   engine oils 51
Fuel economy (see Viscosity)

Gear oil (see Transmissions also)
   choosing 97-107
   multigrade 69, 102
   performance level 66-68
   synthetic 69, 70, 101-107
   service classifications 66-68
   viscosity (see Viscosity)
Gellation index 31, 33, 62
Greases
   calcium-based 78
   clay-based 79
   compatibility/mixing 78, 79
   complex 79
   consistency 79, 80
   performance specification 80
   lithium-based 78
   sodium-based 78

Hypoid gears 39, 66

ILSAC 19, 55

JAMA 55

Lead antiknock 15
Limited slip differentials 71, 104

Index

MIL-L-2105 66
Metal deactivators/passivators 36, 46
Mineral oil 7
Molybdenum disulfide 77
Multigrade viscosity
    engine oils 12, 28, 43, 44, 61
    gear oils 69, 102
Mixed lubrication 37-39, 46

Naphthenic oil 8
Newtonian liquid 42
NLGI 79, 80

Oxidation of oil 32, 45
Overdrives 71, 109

Paraffinic oil 8
Phosphorus 50-51, 84
Polar base oil 47
Poisoning of catalysts 15, 17, 50, 56, 59, 115
Pour point 33
Pour point depressants 33

Running in 49, 118
Rust inhibitor 35

SAE J300 viscosity grades for engine oils 26-27, 61
SAE J306 viscosity grades for gear oils 64-66
Seal swell additive 36
Shear stability 43-45, 62
Shear stabilty index 43

Spiral bevel gears 39
Stable Pour Point 33
Stribeck parameter 37, 38, 48
Synthetic oil – definition 22, 23, 85
    engine oil (see Engine oil – synthetic)
    gear oil (see Gear oil – synthetic)
    use (engine) 119, 120
    viscosity change 29, 45
TBN (total base number) 33
Transmissions (see Gear oil) 63
    automatic (see Automatic transmissions)
Transmission oil (see Gear oil or ATF)

Viscosity – definition 23-25
    effect of temperature on 28-31
    influence on fuel economy 31
    multigrade (see Multigrade viscosity)
    of engine oils 26-27, 61
    of gear oils 64-66
VI (viscosity index) 30
VII (viscosity index improver) 40-45

W or winter viscosity grades
    engine oils 26-28
    gear oils 64-65
Wax in oils 30, 33-34
Wheel bearing lubrication 76-80, 109
Worm drives 99

ZDDP (see ZDTP)
ZDTP (Zinc dithiophosphate) 17, 47, 49, 50, 84, 114-118, 123
Zinc (see ZDTP and Phosphorus)

Also from Veloce –

ISBN: 978-1-845843-51-9 • Paperback • 21x14.8cm • £9.99 UK/$19.95 USA • 96 pages • 32 colour pictures

This book describes in a clear, friendly manner everything today's driver needs to know about choosing and using a car in an economical and eco-efficient way. Includes helpful information on alternative fuels, hybrid powertrains, and much more.

For more info on Veloce titles, visit our website at www.veloce.co.uk • email: info@veloce.co.uk • Tel: +44(0)1305 260068 • prices subject to change, p&p extra